Oyster Cans

HOW TO OBTAIN GOOD OYSTERS
...AND...
HOW TO COOK THEM.

LOUIS GREBB,
OYSTER AND FRUIT PACKER,
MONTFORD AVENUE.
2353-2355-2357-2359 BOSTON STREET,
BALTIMORE, MD.

Jim & Vivian Karsnitz

77 Lower Valley Road, Atglen, PA 19310

DEDICATED
TO
Kathi and Craig
Barb and Jim
Tristan, Ross, Michael and Patrick

Acknowledgements

In acquiring material for this book, we have become indebted to dozens of people. Some opened up their collections to photography and others opened up their minds, providing us with new information. The result of many years of collecting and experience, they gave freely of their special resources. It would be impossible to list everyone and we hesitate to list any, because we must then draw a line. There were however, a few who provided valuable information and went the extra mile in accommodating our questions. R. Lee Burton, Jr., Joseph Davis Jr., Bill and Steve Dorrell, Henry A. Fleckenstein, Jr., J.T. Holleman, Douglas Huether, Frank Hurley, William F. Muckenfuss, Jr., Edmund A. Nelson, Ron Newcomb, Art Oertel, Mike Pinder, Carlton Riggin, T. King Smith, David R. Torre and Cranston Tyler all provided information on oyster cans and the canning industry. Michael Murphy, Harold Reist and Diona M. Snively gave us technical assistance. Our thanks for permission to photograph their collections go to Harry Derstler, L. Craig Hasslinger, and Oscar and Evelyn Schabb. To each we are grateful, for their help has made this effort even more valuable for the readers.

Title page:
Left:
Louis Grebb, Baltimore, MD***
Recipe Booklet
Ronald L. Newcomb Collection

Right:
Rock Hall Clam & Oyster Co., Rock Hall, MD***
Markos Brand, Gallon, MD 68
Carlton and Mary Riggin Collection

Designed by Bonnie Hensley

Library of Congress Catalog Number: 93-85226.

Printed in the United States of America.
ISBN: 0-88740-462-6

Published by Schiffer Publishing, Ltd.
77 Lower Valley Road
Atglen, PA 19310
Please write for a free catalog.
This book may be purchased from the publisher.
Please include $2.95 postage.
Try your bookstore first.

We are interested in hearing from authors
with book ideas on related subjects.

Contents

"Song of the Tongs" Poem**
Joe Sechrist Collection

Foreword

As our enthusiasm for sporting antiques led us into this field of more specialized interest, this book project grew into much more than we originally planned. Cans and tools of the oyster trade are used by some decoy collectors as "go withs," spicing up the display of their collections, and this is what we intended to document. As we began to compile information, we learned much more about oysters and the items associated with them.

We have always enjoyed eating oysters in all the varied methods of their preparation. As we learned more about the industry and its associated objects, we grew to appreciate the oyster even more. Hopefully, the readers also will gain some of our enthusiasm. This is another fascinating area for extended research and collecting pleasure. We invite correspondence on the subject, addressed through the publisher, and look forward to future contacts with kindred souls.

Many of these cans can be found in several sizes. Where there is a photo of a gallon size shown, that does not mean it cannot also be found in quart and pint sizes. To keep the book a manageable size, generally only one example of each design has been included.

The J.M. Clayton Co., Cambridge, MD**
Epicure Brand, Gallon, MD 113
Carlton and Mary Riggin Collection

Values

The value of an item is always a subjective judgment. While most people want to know the values of their possessions, finding an accurate number is often elusive. The quoted figure reflects the judgment of the writer and is not necessarily a market consensus. Also, over time, all values are rendered obsolete.

In light of these problems, the authors have chosen to use a rarity system of rating the items. If the reader feels the need for monetary values, a price range can be assigned to each degree of rarity. The result will be as close to reality as any other way of determining value, and the assigned values can be adjusted for new market conditions.

Each chapter in this book -cans, knives, advertising and other items, must have different values. The item's condition is a key to its value, as damaged, repaired or other imperfections seriously affect price. Where only a few like items are available, or for one-of-a-kind items, the values are widely variable.

In this book, the following rarity designations apply:
* COMMON
** SCARCE
*** RARE
**** VERY RARE
***** ONLY A FEW KNOWN OR ONE OF A KIND

Cans and Shipping Containers Salvaged From Side
Wheel River Steamboat **Arabia**
Courtesy of the Arabia Steamboat Museum

Chapter I

Oyster Containers, Cans, Jugs and Bottles

The preservation of food was a persistent challenge. The French government of Napoleon was especially interested in preserving its soldiers' food in the field. Although the soldiers were well trained, they could fight well only if they were in good condition. In 1795, Napoleon offered a prize of 12,000 francs for a reliable method of preserving food.[1] In 1809, Nicolas Appert won the prize by developing a method of sterilizing food in glass containers. When he wrote a book a year later outlining his method, the world learned how he preserved foods. Other experimenters, like Peter Durand, introduced Appert's method to England and later, in 1818, to America. Because glass containers often broke while being transported in wagons, metal containers were tried. Hand made tin containers called canisters, later shortened to cans, proved better suited for commercial purposes.

In America by 1819, Thomas Kensett I and Ezra Daggett canned a few oysters in New York City. By 1839 most glass containers were replaced by tin-plated cans, first in rectangular shapes and then round ones. The following year oyster canning began in Baltimore, Maryland. Oysters were then hauled west over the National Pike in Conestoga wagons as far as Pittsburgh.

Oysters were shipped throughout the American midwest in cans packed in wooden cases. A side wheel river steamboat, the *Arabia*, was sunk in the Missouri River near Kansas City in 1856 with 220 tons of cargo destined for the frontier. Recently, the steamer was salvaged and its cargo placed on display in a museum in Kansas City. Among the cargo were wooden boxes containing canned oysters in two sizes. The Baltimore packer Price & Littig had shipped them to consignees in the west. The boxes have survived in surprisingly good condition, the labels were missing, but the cans were fine and the stenciling on the cases is easily read.[1]

Oyster Shucking**
Harpers Weekly March 1872

From Baltimore, oyster canning moved south and west. Oysters were canned in Apalachicola, Florida in 1884, Brunswick, Georgia about 1886, and Biloxi, Mississippi about 1915. Biloxi surpassed Baltimore as an oyster canning center in the first half of the twentieth century and packed more oysters than all the other states. On the west coast, Puget Sound oysters were canned by 1931.[2]

Cans first were made by hand, cutting tin bodies that were bent around a cylindrical mold and the seam soldered. Tops and bottoms were also cut and soldered to the body. A cap hole was left in the top through which oysters were inserted. This was closed by soldering a cap over the hole after filling. Sixty canisters a day was the maximum output of an expert craftsman. This same can was also used by the vegetable and fruit canners.

[1] On display at the *Arabia* Steamboat Museum, Kansas City, Missouri.

[2] Much of the historical information was gathered by Earl Chapin May and R. Lee Burton, Jr. whose pioneering work documented the industry's beginnings. Their books are listed in the bibliography.

Filling with Liquor**
Harpers Weekly March 1872

Oyster Canning**
Harpers Weekly March 1872

The Bath**
HarpersWeekly March 1872

Cans first were made by hand, cutting tin bodies that were bent around a cylindrical mold and the seam soldered. Tops and bottoms were also cut and soldered to the body. A cap hole was left in the top through which oysters were inserted. This was closed by soldering a cap over the hole after filling. Sixty canisters a day was the maximum output of an expert craftsman. This same can was also used by the vegetable and fruit canners.

Steerhead Can showing Fill Hole****

Soldering the Cans**
Harpers Weekly March 1872

When Thomas Kensett II moved to Baltimore in 1849, oyster canning began in earnest. Baltimore was ideal because it was close to the oyster beds and labor was plentiful. A pattern developed to can oysters in the winter and fruits and vegetables in the summer, providing year-round business. By 1880 Baltimore was the center of the canning industry, and a supply center for cans, labels and shipping containers.

Pendant from Baltimore, MD Veterans Convention***
Indicating Universal use of Oyster Motif in Chesapeake Culture
Ronald L. Newcomb Collection

An interesting marketing note is the use of the term "Cove Oysters" in the packaging. Many people believe the term meant from a cove in the Chesapeake Bay, but it was begun in an entirely different way. Two "oyster bars" were located near each other in the 1870s on a Baltimore street called Cove Street. One bar sold fresh raw oysters while the other sold oysters in hermetically sealed cans. The dealer of the raw oysters objected to the dealer across the street calling his cans "Fresh Chesapeake Oysters," so he started calling the canned ones "Fresh Cove Oysters." In a short time, the name came to mean any canned Baltimore oyster. A fire in 1904 wiped out Cove Street and many old oyster establishments.

Set of Half, Gallon, Three and Five Gallon Oyster Cans****

Over the years better methods of making cans and preserving were developed. Allen Taylor patented a machine to stamp cans with extension edges in 1847, and in 1849 Henry Evans invented a "pendulum press" for semi-automatically making can tops and bottoms. In the same year, William Numsen adopted the "combination die" for can making. In 1851 Louis McMurray used the term "hermetically sealed oysters" and his product traveled around the world. Isaac Solomon, in 1861, added calcium chloride to the water in the cooking kettles, raising the boiling point to 240 degrees, making for more complete sterilization.

An Oyster Speculator Testing a Cargo**
Harpers Weekly March 1889

Oystering on Long Island**
Harpers Weekly October 1886

Oyster Roast at Pimlico Race Track,**
Baltimore, MD
Leslies Illustrated February 1894

New ways of applying solder to can making and improved machinery enabled the start of assembly-line production. All this technology put employees out of work and violence resulted. Machinery was broken and owners' homes were damaged. Technological unemployment is not a new phenomenon.

The packing of oysters for shipment was carried out by many different companies under variable sanitation conditions. Not much was known about sanitary measures, and little thought was given to the possibility of infectious conditions in the plant or during shipment. Packers believed their product was as good as any food in this regard, but outbreaks of typhoid fever and gastro-intestinal disorders occurred. As a result, health authorities began looking into the problem and packers began to lose business because oysters were blamed on many outbreaks. Repeated discussions, sometimes heated, followed and the federal government was asked to help.

In 1909, the Department of Agriculture issued Pure Food Decision No. 110, prohibiting certain practices and requiring sanitary measures in the packing houses. In 1925 more problems came with the outbreak of typhoid fever traced to raw polluted shellfish. As a result, each state was required to regulate the industry with regards to sanitation. Numbered certificates were issued to each packing house enabling the authorities to trace all shipments to a specific packer. Each packing plant was assigned a number regardless of ownership, thus one company could have several numbers. If the company contracted with a second company to pack under the first company's brand, the number on the can would be the second company's number.

Numbers were assigned as plants were checked or applications made, so numbers are no indication of age. In addition, one company's can may have different numbers,

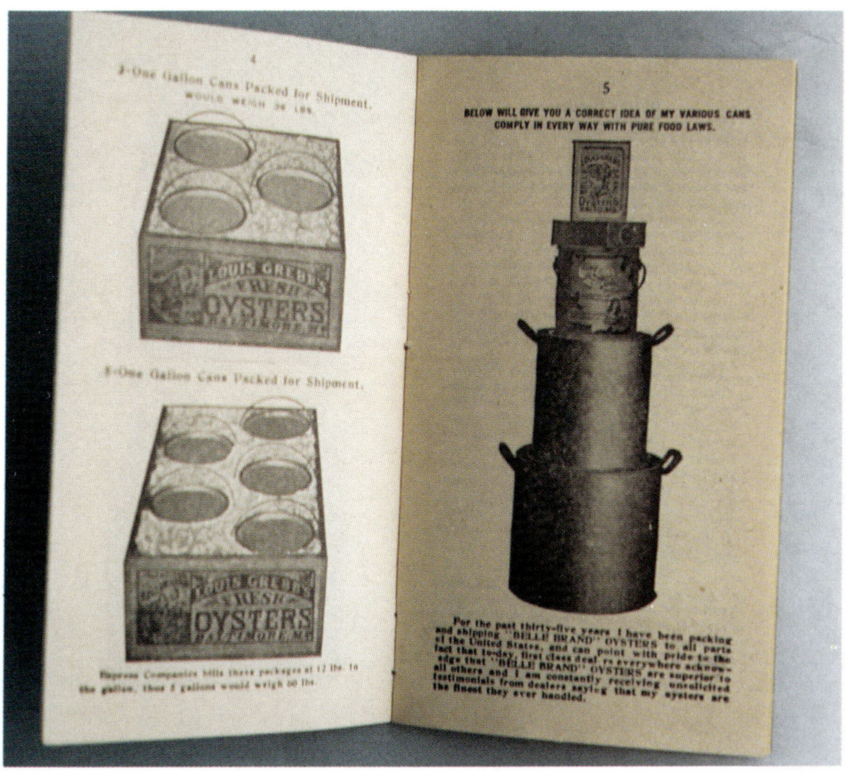

Louis Grebb Catalog Showing***
Arrangement of Oyster Cans for Shipping
Ronald L. Newcomb Collection

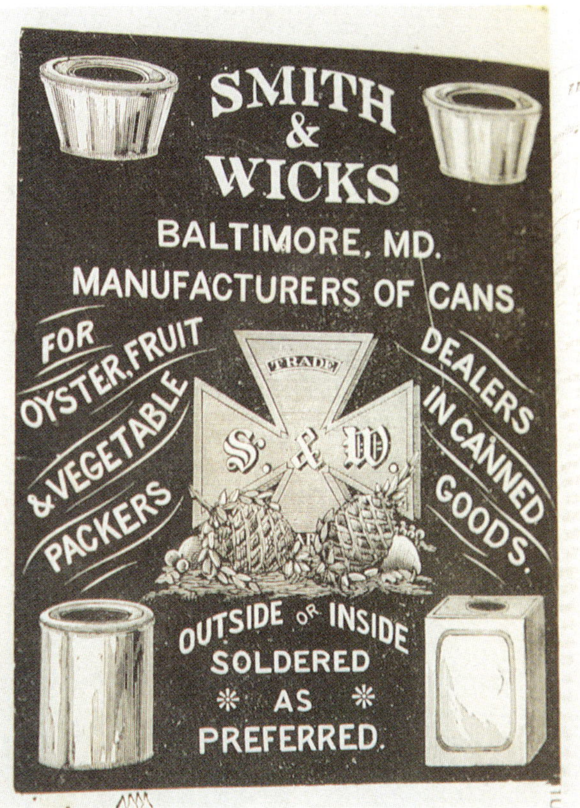

Advertisement of Smith and Wicks Can Company,**
Baltimore, MD

depending on where it was filled. Companies that went out of business might have their number reassigned to a new company, further complicating matters.

Some dating is possible using postal zones and zip codes. The one and two digit zone code system used by the large post offices began on May 1, 1943 and the Zone Improvement Plan (ZIP) code using five digits started July 1, 1963. This date cannot be absolute, since any packer with a supply of cans used them up before ordering new ones incorporating ZIP codes.

Oyster can tops are marked either Extra-Counts, Counts, Extra Selects, Selects, Extra-Standards, and Standards. This means the number of oysters in a gallon approximately as follows:

Extra Counts	under 90 per gallon
Counts	from 90 to 150 per gallon
Extra Select	from 150 to 200 per gallon
Select	from 200 to 240 per gallon
Extra Standards	from 240 to 280 per gallon
Standards	from 280 to 400 per gallon

As in other industries during this period mergers of small companies into industrial giants occurred. The American Tinplate Company became the American Sheet and Tinplate Company in the period 1888-89. This company, along with others, was merged into the American Can Company in 1901, beginning the company as it is known today. Similarly, in 1905, Continental Can Company began operations through the consolidation of smaller companies including Southern Can Company of Baltimore. Some cans were made by the Metal Packaging Corp. of Baltimore. These cans are marked M. P. Co. Most of the can manufacturers were located in Baltimore. Those located around the Chesapeake Bay merged with Baltimore companies. Cans were shipped to the packers first by steamship and later by railroad or truck. Continental Can Company, American Can Company, National Can Company, Steeltin Can Company and Independent Can Company made the majority of oyster cans. Steeltin Can Company and Independent Can Company were locally owned, and National was merged with American.

HAVRE DE GRACE REPUBLICAN ALMANAC.

NEW YORK HOME MADE BREAD,

MANUFACTURED BY

JNO. W. CRUETT,

41, 43 & 45 S. Eden Street, BALTIMORE, MD.

The proprietor gives his personal attention to the management of his business, and hence the superiority of his production.

He makes a specialty of supplying Harford County Canning establishments, and guarantees satisfaction, as his numerous patrons in Harford will cheerfully attest.

His facilities for supplying hotels, restaurants, railroad and other contractors and dealers, are unequalled, and he only solicits a trial order to prove the truth of what is herein stated.

☞ *A personal call is solicited, or all inquiries cheerfully answered by addressing the proprietor as above.*

J. H. C. Thirlkel,

Manufacturer of Oyster and Fruit Cans,

Also of Paint, Putty, Lard and Oil Cans, Fruit Butter, Mince Meat and Jelly Buckets.

—SOLDER *and* CASES *furnished at lowest market price.*—

79 & 81 EAST PRATT STREET,

Advertisement of J. H. C. Thirlkel Can Company,**
Baltimore, MD

Advertisement of Johnson-Morse Can Co.,**
Wheeling, WV

THE Canning TRADE

THE "BOYER" CANS

HAVE NO EQUAL

They are heavier than others.
Stronger, with fewer Leaks.
The cleanest ever produced.

OPEN TOP (SANITARY) CANS OF FINEST QUALITY

Latest Improved Closing Machines

Favorable Terms — Satisfaction Assured

ALSO

PACKERS' CANS—all sizes up to No. 10

INCLUDING JERSEYS AND NO. 2½.

SOLDER HEMMED CAPS

PERFECTLY MADE WITH SOLID RINGS OF GOOD SOLDER, AND PLENTY OF IT.

Also Friction Top Cans for

OYSTERS, CRAB MEAT, SYRUP, Etc., and WAX TOPS.

FINE EQUIPMENT AND LARGE CAPACITY. BEST SHIPPING FACILITIES.
RAILROAD SWITCH AND WATER.

We Solicit Your Orders and Guarantee Satisfaction

W. W. BOYER & CO.

2327 Boston Street — BALTIMORE, MD.

Advertisement of W.W.Boyer & Co.,**
Baltimore, MD

"Southern" Cans for 1914

Purchase, Pack and Prosper

Made in Baltimore

12 YEARS OF UNINTERRUPTED SUCCESS

On this the first occasion of the National Canners' convening in Baltimore, we have twel most successful years upon which to look back, and although during this period both the sellin and manufacturing methods of the business have been revolutionized our steady progress prove we have been—and still are—abreast of the times. Our present modern plant employing s hundred hands is the third which we have occupied since our incorporation, the frequent chang of location being made necessary from our phenomenal growth.

CAP HOLE CANS

To our old customers, who have packed all varieties of fruits and vegetables in "Southern" cans, comment on their high quality is unnecessary. Of others, not so familiar we simply ask an opportunity of demonstrating our claims. The fit of our hemmed caps and the quality and quantity of the solder thereon are not the least of their good features.

THE GENERAL LINE

Few Can manufacturers have at tempted to make Packers cans an those for special purposes in the sam plant. We have been successfu through a well trained force and amp storage. Our claim to make the mo diversified line of any Independent Ca Company in the country is yet to b disproved. It comprises cans for Pain Putty, Baking Powder, Coffee, Confe tionery, Oysters, Crab Meat, etc., et Also the Wax Top and Friction T Styles and Lard Pails.

SANITARY (OPEN TOP) CANS

Don't conclude that simply because we have been in this particular line but one year we are incompetent. True, our experience is short but most satisfactory. We are frank to say we have profited by the errors of others. "Patience" was our motto while machines and methods were being perfected, and we are now reaping the benefits—which will be shared with our customers.

LITHOGRAPHING and METAL SIGN DEPARTMENT

In this department we not o decorate packages in an economical a artistic manner, but here we do all c lacquering for the inside of food co tainers when required. Metal Signs advertising purposes of both simple a elaborate design form no small part this business.

UNSURPASSED SHIPPING FACILITIES

The recent introduction of a switch from the Pennsylvani Railroad into our plant, in addition to similar accommodation already provided by the Baltimore & Ohio Railroad, togeth with access to the water front makes conditions ideal. Spac for eighteen cars is now available on our own trackage.

Old and prospective customers will be cheerfully granted the opportunity to inspect our plant during their stay at the Canners' Convention, February 2nd to 7th, 1914.

E. EVERETT GIBBS, President
LEONARD BURBANK, V.-P. and
J. OLIVER SELBY, Secretary
JOS. M. WIEST, Sales Mgr.
FRED H. CEARFOSS, Mgr. Sign D
HARRY E. PACKARD, Asst. Sale
JAS. A. COLBERT, Asst. Sales Mg
JOSEPH G. FICK, Traffic Mgr.

Southern Can Co.

717 South Wolfe Street — Baltimore, Maryland

Advertisement of Southern Can Co.,**
Baltimore, MD

THE Canning TRADE

Syracuse Chicago New York Baltimore Canonsburg

CONTINENTAL CAN COMPANY

Incorporated

TO OUR FRIENDS AND PATRONS:

We invite your especial attention to our Exhibit at the National Convention, to be held at Baltimore in February, of our Latest Improved High Speed Automatic Closing Machines, which are readily interchangeable for the closing of different sizes of Open Top Cans.

These are the only machines now on the market containing all of these very important features.

We will be very glad to have you give our machines your most critical inspection.

Yours very truly,

CONTINENTAL CAN COMPANY, Inc.

T. G. Cranwell, President.

Advertisement of Continental Can Co.,**
New York

After bottles were phased out in favor of tin-plated cans around 1839, the first cans had a fill hole in the top. These were replaced by the friction tops, usually in pint and gallon sizes. Some half-gallon, three and five gallon cans were also in use with information embossed into the bodies of all these cans. Included was the capacity of the can and the name of the packer. The half-gallons and gallons had bail handles and all were very plain "tin cans." Can lids were also embossed with the packer's name and the size of the oysters packed in the can. It is rare today to find a can with the original top. In 1880 George Seib introduced a lithographed label and when lithographed oyster cans came into use in the 1920s and 1930s, the colorful cans we know today

Capping Machine**
Mike and Eva Pinder Collection

THE Canning TRADE

HOLE AND CAP
AND
SANITARY CANS

No matter what style of can is used it is a matter of common business prudence to use the best made of the desired type.

* * * Service coupled with quality is an equal consideration.

Service means "a whole lot" embracing as it does every detail co-incident with the handling of the business.

It means prompt shipments—cans when you want them.

Shortest mileage to your factory—low freight rates and quick time.

A proper personal consideration of needs before and after cans are delivered.

It is on the basis of quality and service that the American Can Company solicits your business.

AMERICAN CAN COMPANY

Chicago
Baltimore

NEW YORK
Rochester, N. Y.

San Francisco
Portland, Ore.

Advertisement of American Can Co.,**
New York

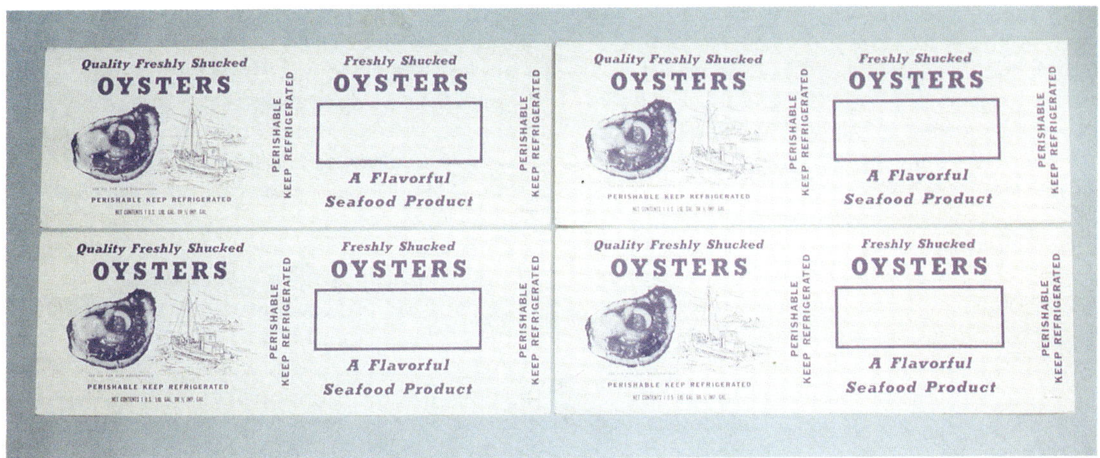

Sheet of Four Can Bodies*

began to be made. Bail handles were phased out in the 1920s and 1930s because of the extra cost and with the start of World War II, steel restrictions finally eliminated them.

The gallon-size oyster can was modeled after paint cans with bail handles which had a triple plug top with three rolls to make this seal. The oyster can, however, had only a single friction or plug top closure. Pint cans were made the same way but without the bail handles.

Competition was keen and new ideas were constantly being developed to gain a competitive advantage. Edmund A. Nelson, who left his father's business, White and Nelson Seafood of Cambridge, Maryland, in 1956 to sell cans for Steel & Tin Products Co. of Baltimore, has related many of the following interesting facts about the can business. Steel & Tin Products Co. organized in 1912, made the first crabmeat can about 1930 and oyster cans a few years later. In the late 1930s or early 1940s, the open top pint can, requiring the lid to be attached by machine, came into use. The half-pint can followed a few years later and the twelve ounce a few years after that. In the early 1950s the window top was developed in response to a request by people at the A & P Food Stores to enable customers to both see the product and eliminate glass containers. About 1970, plastic containers began to replace metal ones for all oyster containers. By 1985 the change-over was almost complete and plastic is the container of choice today. In addition to being lighter and cheaper, plastic containers can be nested for shipping to the packer, thus saving space and shipping costs.

The packers who created their own designs on cans had to place a minimum order of twelve hundred cans. Later this minimum was increased to two thousand cans. If a packer could not use this quantity, stock or generic cans were also available with standardized sailboats, oyster designs or plain lettering. All the essential information was lithographed on the can. Some of the larger packers, distributors and large chain stores would ship their lithographed cans to other packing houses to be filled. The regulations required the name and certificate number of the plant that packed the oysters to be on the container prior to shipment from the manufacturer. In this case, the larger packers certificate number was left off the lithography and the name and number of the plant that packed the oysters was embossed in the can; the packer could not add them himself. This procedure also required the wording on the can to be changed from "Packed By-" to "Packed For-" or "Distributed By-." Thus, every can delivered to the consumer carried a packer's number. Gallon cans were shipped 36 to a Kraft bag or 51 in a corrugated carton, depending on the size of the order. The Steel & Tin Products Co. name was changed to Steeltin Can Corp. about 1970. The logo for Steel and Tin Products was STP with the T & P inside the loops of the S, used until 1970. For Steeltin Can Corp. "Steeltin" was used from 1970 to 1980, at present the logo is SCC, with the S and the CC being formed by the stylized ends of two cans. Steeltin was located on President Street until 1956 when it moved to 8301 Pulaski Highway.

Mr. Douglas Huether, president, relates that Independent Can Company originally was located at Howard and Ostend Streets in Baltimore. From 1949 to 1952 it was located at 237 President Street and from 1952 to 1990 on South Lakewood Avenue in the Canton section of Baltimore. In 1990 Independent moved into modern facilities in Belcamp, Maryland. The original logo, used from 1929 to 1949 was "Independent Can Co. Baltimore, MD". In 1950 it was shortened to "Independent Can Co., Balt. MD".

Chesapeake Can Company of Crisfield, Maryland, operated by Harry Wells, manufactured cans in Crisfield until early in World War II when limited availability of tin mill products caused it to close. He became a distributor of Independent Can Company products, at that time maintaining a warehouse there, closing shortly after the wars end.

Independent Can Co. was incorporated in 1928 and by 1929 was making oyster cans. Before the second World War the steel was hot dipped to coat it with tin. During the war they made bonderized black plate cans from chemically treated steel. After the war, they switched back to electroplated tin and today chrome oxide plating is used.

Plug tops were used for pints and quarts by all companies except American and Continental can companies. They leased sealing machines to their customers enabling them to use open top cans. A 1950 court judgment in San Francisco extended the use of sealing machines to everyone and the open top cans came into general use.

In 1962, experiments with plastic containers began and by 1975 tin and plastic were used equally. Today, K-resin containers as clear as glass and far less breakable are used with only a very small number of packers still using the tin cans, primarily 8 and 12 ounce sizes because chain stores want a lower unit cost.

In 1949 the Parker family, owners, sold their interest to Harry L. Huether. He operated a tinplate warehouse and had a close contact with Independent Can Co. The plant manager was Gordon Tyler, hired from Crown, Cork and Seal Co. Douglas Huether is carrying on today with his son, Rick, and his son-in-law George McClelland, also active in the business.

Many of the designs on oyster cans originated with company sales representatives and packing house owners. Commercial artists translated the ideas into finished designs. The lithographer printed the design on metal strips two bodies wide and two long, four can bodies to each sheet. The sheets were then delivered to the can manufacturer who cut the sheets apart and formed the body and attached a top rim and a bottom. Designs were more intricate on steel cans using a 120-line screen than on today's plastic containers on which 80 to 100-line screens are used. Colors were printed separately on tin while all colors are printed simultaneously on plastic.

T. King Smith, a commercial artist who designed most of the oyster and crab meat cans still designs many containers. He relates that most of his customers asked for a boat, lighthouse, oyster or river to be incorporated into the design, but left the rest up to him. Sometimes he would get a picture of the packer to include and there are unusual stories about how some designs came to be. The majority of cans were printed in one or two colors, black, dark blue, black and red, or dark blue and red. Very few cans were printed in more than two colors because it was too costly.

The packer for the Stork Brand of Thomas and Thompson, later H.S. Thompson and Son, was thinking of a design for his can while driving home one day. When he saw a heron standing in water, he thought it was a stork and liked it. The Stork Brand was born. Another idea came from a nickel. The Honga Brand of White and Nelson of Cambridge, Maryland used the Indian head from the old nickel as its inspiration. This company began on Hoopers Island, Maryland in 1912, and has one of the more attractive cans. It would be interesting to know how many other brands were conceived this way.

Can collectors often find the cans they acquire have been stored near salt water in warehouses. The cans are often rusty and covered with accumulated dust. Most want to remove the rust and dirt to preserve them from additional damage. A mild soap and warm water will remove the grime. Soaking the can in lemon juice or similar liquid will remove the rust. A product used to restore and brighten automobile finishes will do the same to metal cans. One product, Meguiar's Car Cleaner-Wax sold by Western Auto Stores, is recommended by some collectors although any quality product should work. Before doing anything, collectors should experiment on an inconspicuous area, to find out what works for them. After the can is cleaned a good grade of clear paste wax will provide continuing protection.

In addition to metal cans, oysters were also packed in stoneware and glass containers. Stoneware was used primarily from Boston, Massachusetts northward to Maine, and glass bottles, similar to milk bottles, generally were used from Boston to southern New England; examples of both are found in Boston. These containers are very rare and difficult to find.

Oyster cans can be found in various categories including specialized geographical areas, types of cans, individual companies or special graphics. Cans of individual packers sometimes changed, showing a sequence of design or ownership. The R.E. Roberts Co. of Baltimore changed their "Maryland Beauty" whose picture was redrawn to keep up with fashion. The "Stork Brand" is an example of ownership change as well as style change. The design changed once and later the same design had the name of the packer changed. It is necessary to look closely at each can because of these variations, much as stamp or coin collectors do. Shown are over eight hundred color pictures of cans from around the country and an additional two hundred pictures of related items.

In the captions accompanying the pictures the name of the packer, brand name, can size and packers number will be listed in that order, if available. Some embossed cans and bottles have had the raised letters highlighted with paint, they are normally not found this way.

A companion volume by the authors shows oyster plates and silverware and discusses oyster plate manufacturers and design patents. The plates are a large and varied group made from porcelain and majolica china in an endless parade of designs and colors, manufactured in both the United States and Europe during Victorian times.

Commercial Artist Designing Oyster Cans

MARYLAND

Wm. D. Gude & Co., Baltimore, MD****
Premium Brand, Gallon

Wm. D. Gude & Co., Baltimore, MD****
Premium Brand, Gallon
Mike and Eva Pinder Collection

J.H. Collison Co., Baltimore, MD****
QualiT Brand, Gallon
Ronald L. Newcomb Collection

J.H. Collison Co., Baltimore, MD****
QualiT Brand, Tall Quart
Bill and Steve Dorrell Collection

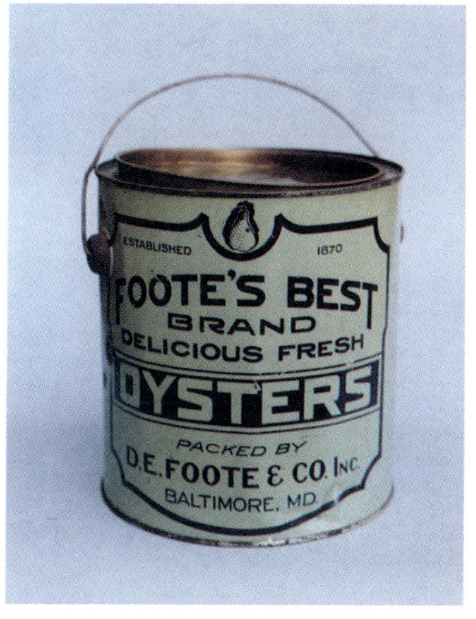

D.E. Foote & Co., Baltimore, MD****
Foote's Best, Gallon
George and Betty Juergens Collection

R.J. McAllister & Co., Baltimore, MD***
Gallon, MD 43

J.Langrall & Bro., Baltimore, MD**
Gallon
Carlton and Mary Riggin Collection

Louis Grebb, Baltimore, MD****
Belle Brand, Quart
Frank Speal, Jr. Collection

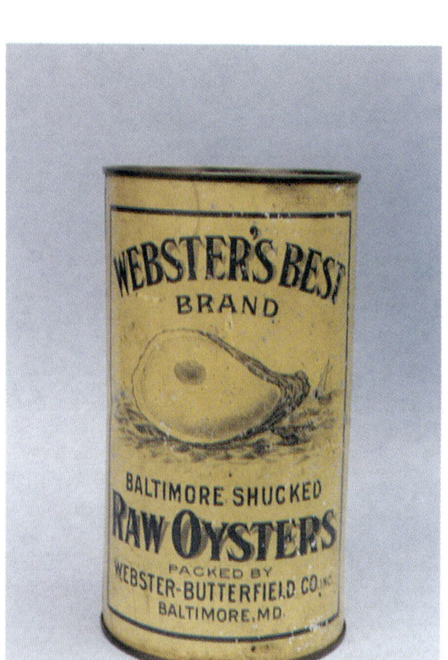

Webster-Butterfield Co., Baltimore, MD****
Webster's Best, Tall Quart
Ronald L. Newcomb Collection

F.C. Bower & Co., Baltimore, MD****
Regina Brand, Half Gallon
Ronald L. Newcomb Collection

Louis Grebb, Baltimore, MD****
Belle Brand, Gallon
Ronald L. Newcomb Collection

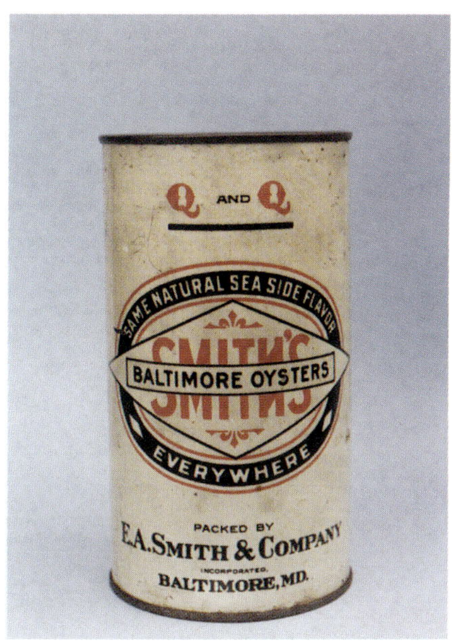

E.A. Smith & Company, Baltimore, MD****
Smith's Baltimore Oysters, Tall Quart
Ronald L. Newcomb Collection

Harry Rochester Co., Baltimore, MD***
Pint, MD 44
Ronald L. Newcomb Collection

E.A. Smith & Company, Baltimore, MD****
Smith's Baltimore Oysters, Gallon
Randy Shreck Collection

E.A. Smith & Company, Baltimore, MD****
Smith's Baltimore, Oysters, Tall Quart

Harry Rochester Co., Baltimore, MD****
Maryland's Pride, Pint
Ronald L. Newcomb Collection

Ocean Seafood Co., Baltimore, MD****
Purity Brand, Gallon
Randy Shreck Collection

F. Bonhage & Co., Baltimore, MD*****
Elk Brand, Quart
Carlton and Mary Riggin Collection

C.H. Lighthiser, Baltimore, MD****
Elephant Brand, Tall Quart
Tyler Campbell Collection

C.H. Lighthiser, Baltimore, MD****
Elephant Brand, Tall Quart
Mike and Eva Pinder Collection

C.H. Lighthiser, Baltimore, MD****
Elephant Brand, Gallon
Ronald L. Newcomb Collection

H. McWilliams & Co., Baltimore, MD****
Express Brand, Gallon, MD 24
Carlton and Mary Riggin Collection

Crescent Sea Food Co., Baltimore, MD****
Crescent Brand, Gallon, MD 26
Carlton and Mary Riggin Collection

H. McWilliams & Co., Baltimore, MD****
Express Brand, Gallon
Ronald L. Newcomb Collection

Wm. Jacobs & Sons, Baltimore, MD****
Chef Brand, Gallon, MD 46
Ronald L. Newcomb Collection

Lord-Mott Co., Baltimore, MD**
Old Reliable Brand, Pint, MD 17
Carlton and Mary Riggin Collection

J.D. Groves & Co., Baltimore, MD****
Pride of the Chesapeake, Tall Quart
Ronald L. Newcomb Collection

Lord-Mott Co., Baltimore, MD****
Old Reliable Brand, Gallon, MD 17
Mike and Eva Pinder Collection

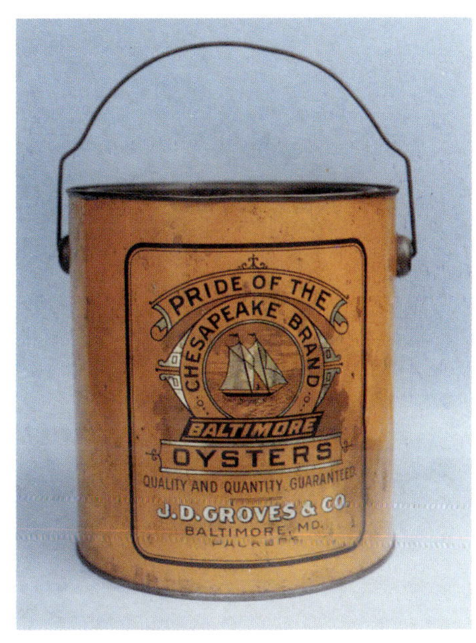

J.D. Groves & Co., Baltimore, MD****
Pride of the Chesapeake Brand, Gallon
Ronald L. Newcomb Collection

W.H. Killian Co., Baltimore, MD****
Sea-L-tite Brand, Gallon, MD 12
Carlton and Mary Riggin Collection

Lord-Mott Co., Baltimore, MD***
Old Reliable Brand, Gallon, MD 17
Ronald L. Newcomb Collection

Wm. J. Nollmeyer, Baltimore, MD***
Gallon
Carlton and Mary Riggin Collection

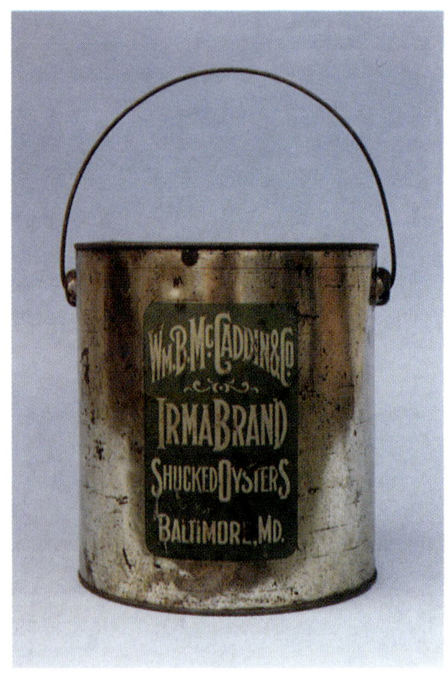

Wm. B. McCaddin & Co., Baltimore, MD****
Irma Brand, Gallon

E.A. Lux Packing Co., Baltimore, MD**
Lux Brand, Pint, MD 168
Carlton and Mary Riggin Collection

John T. McNaney, Baltimore, MD****
McNaney Brand, Gallon, MD 20
Ronald L. Newcomb Collection

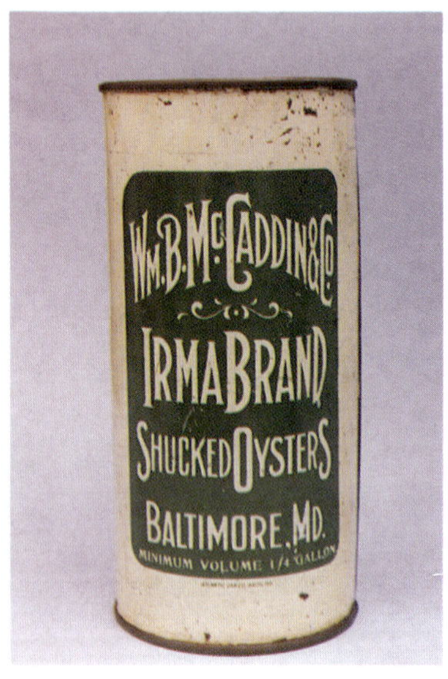

Wm. B. McCaddin & Co., Baltimore, MD****
Irma Brand, ¼ Gallon
Ronald L. Newcomb Collection

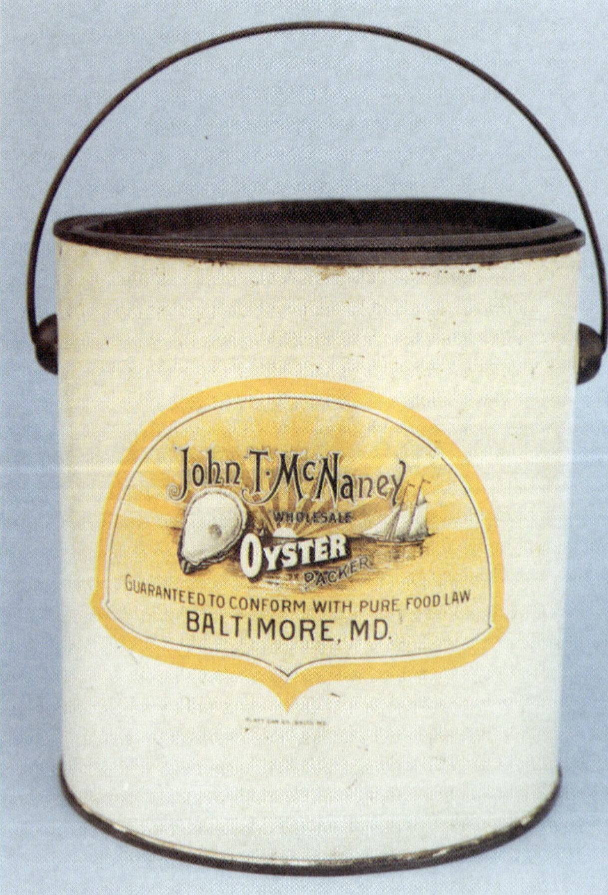

John T. McNaney, Baltimore, MD****
Gallon
Gary and Sharon Campbell Collection

Atlantic Packing Co., Baltimore, MD****
Majestic Brand, Gallon
Mike and Eva Pinder Collection

McNaney Oyster Co., Baltimore, MD****
Superior Brand, Gallon and Pint, MD 14

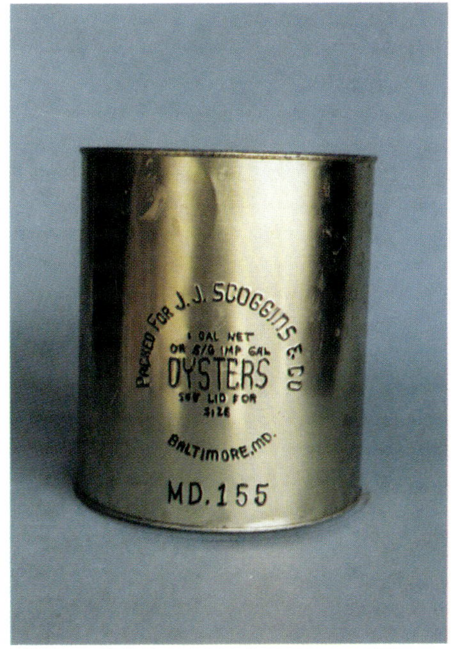

J.J. Scroggins & Co., Baltimore, MD**
Gallon, MD 155
Carlton and Mary Riggin Collection

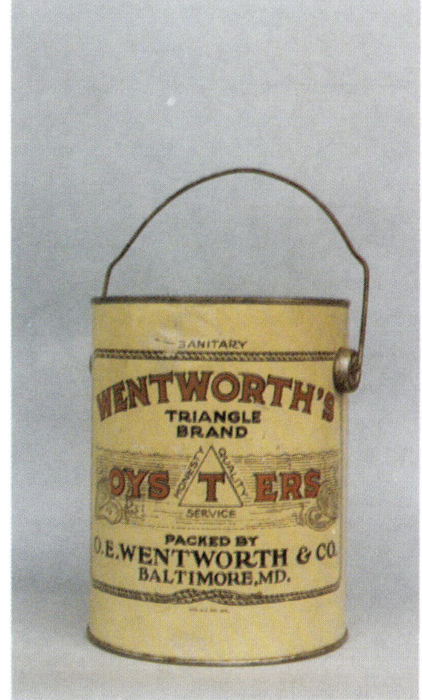

O.E. Wentworth & Co., Baltimore, MD****
Triangle Brand, Gallon, MD 45
Carlton and Mary Riggin Collection

O.E. Wentworth & Co., Baltimore, MD****
Triangle Brand, Gallon, MD 45
Carlton and Mary Riggin Collection

Crisfield Sea Food Co., Baltimore, MD****
Deep River Brand, Gallon, MD 50

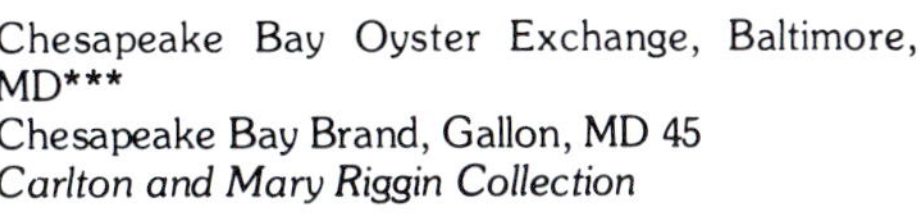

Chesapeake Bay Oyster Exchange, Baltimore, MD***
Chesapeake Bay Brand, Gallon, MD 45
Carlton and Mary Riggin Collection

Wm. Heyser, Baltimore, MD****
Heyser's, Gallon
Ronald L. Newcomb Collection

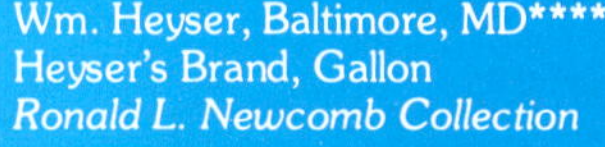
Wm. Heyser, Baltimore, MD****
Heyser's Brand, Gallon
Ronald L. Newcomb Collection

The Wm. Heyser Co., Baltimore, MD****
Heyser's Brand, Gallon
Ronald L. Newcomb Collection

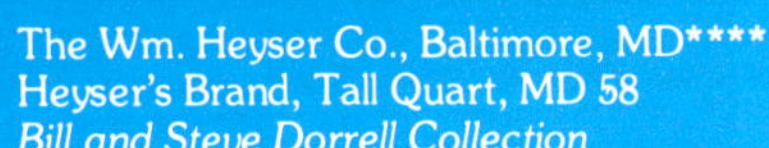
The Wm. Heyser Co., Baltimore, MD****
Heyser's Brand, Tall Quart, MD 58
Bill and Steve Dorrell Collection

C.L. Applegarth Co., Baltimore, MD****
Acme Brand, Gallon, MD 51
Randy Shreck Collection

C.L. Applegarth Co., Baltimore, MD*****
Acme Brand, 1/6 Gallon

C.L. Applegarth Co., Baltimore, MD
Can Lid
Carlton and Mary Riggin Collection

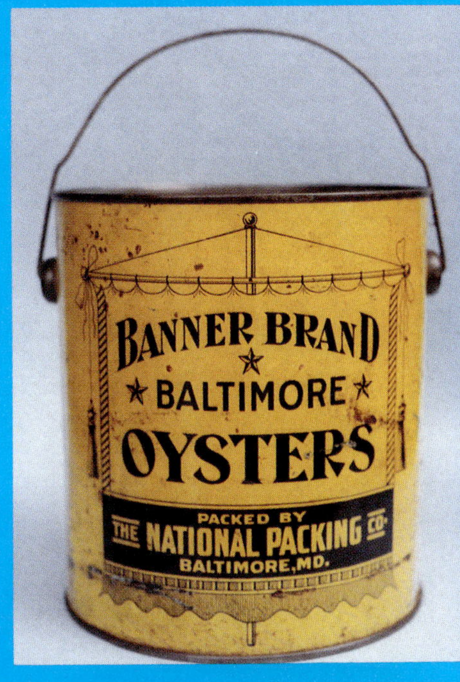

The National Packing Co., Baltimore MD****
Banner Brand, Gallon
Ronald L. Newcomb Collection

Travers Bros. Co., Baltimore, MD****
Blue Cross Brand, Tall Quart
Ronald L. Newcomb Collection

The National Packing Co., Baltimore, MD****
Banner Brand, Tall Quart and Pint, MD 222
Ronald L. Newcomb Collection

C.L. Applegarth Co., Baltimore, MD***
Acme Brand, Pint, MD 51
Randy Shreck Collection

Travers Bros. Co., Baltimore, MD****
Blue Cross Brand, Tall Quart
Ronald L. Newcomb Collection

Travers Bros. Co., Baltimore, MD****
Blue Cross Brand, Gallon
Carlton and Mary Riggin Collection

J.C. Coulbourn Co., Baltimore, MD****
Big "C" Brand, Half Gallon, MD 47
Carlton and Mary Riggin Collection

Chas. W. Reddish, Baltimore, MD**
Gallon
Carlton and Mary Riggin Collection

J.H. White Co., Baltimore, MD****
DeLuxe Brand, Gallon, MD 50
Carlton and Mary Riggin Collection

Chas. W. Reddish, Baltimore, MD****
Tall Quart and Pint
Ronald L. Newcomb Collection

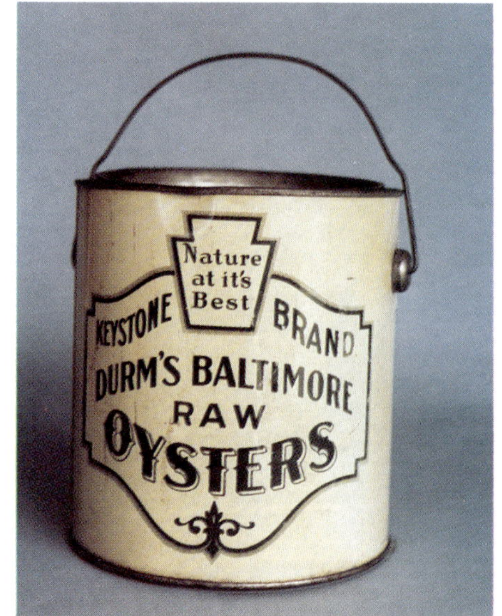

Wm. B. Durm & Co., Baltimore, MD****
Keystone Brand, Gallon, MD 61
Carlton and Mary Riggin Collection

J.H. White Co., Baltimore, MD**
Gallon, MD 50
Carlton and Mary Riggin Collection

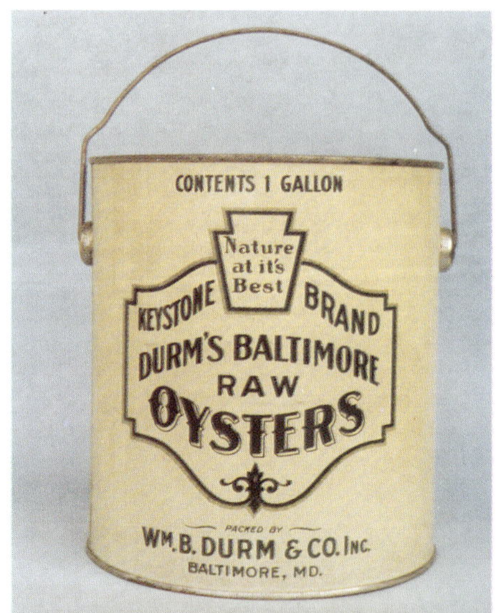

Wm. B. Durm & Co., Baltimore, MD****
Keystone Brand, Gallon, MD 61
Butch and Jackie Cheezum Collection

Southern Sea Food Co., Baltimore, MD****
Double "S" Brand, Quart, MD 24
Ronald L. Newcomb Collection

Jas. Hubbard & Son, Baltimore, MD****
Plymouth Rock Brand, Gallon
Ronald L. Newcomb Collection

J.W. Chew, Baltimore, MD****
Gallon, MD 59
Carlton and Mary Riggin Collection

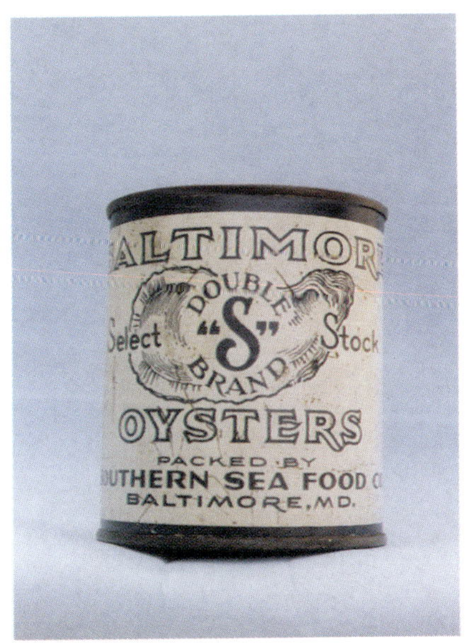

Southern Sea Food Co., Baltimore, MD****
Double "S" Brand, Pint, MD 24
Bill and Steve Dorrell Collection

J.W. Chew, Baltimore, MD****
Gallon

J.W. Chew, Baltimore****
Tall Quart
Courtesy of Black Swan Antiques

Standard Fish & Oyster Co.,
Baltimore, MD****
Plymouth Rock Brand, Tall Quart
Ronald L. Newcomb Collection

Planters Packing Co.,
Baltimore, MD****
Stag Brand, Tall Quart
Butch and Jackie Cheezum Collection

*Planters Packing Co., Baltimore, MD*****
Stag Brand, Tall Quart
Bill and Steve Dorrell Collection

Planters Trading Co., Baltimore, MD****
Stag Brand, Gallon
George and Betty Juergens Collection

Planters Packing Co., Baltimore, MD**
Stag Brand, Gallon
Carlton and Mary Riggin Collection

Eagle Packing Co., Baltimore, MD****
Eagle Brand, Gallon
Ronald L. Newcomb Collection

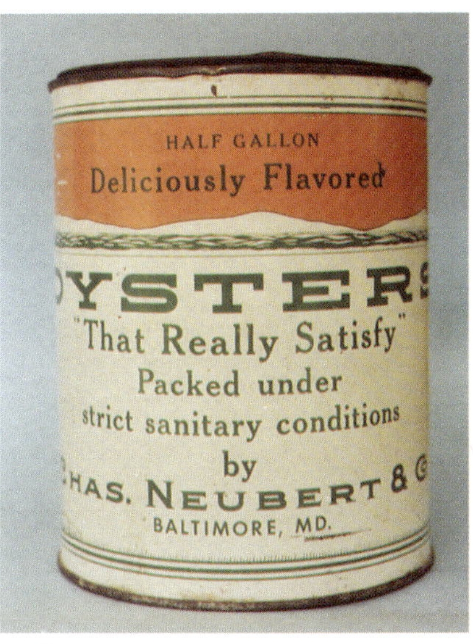

Top left:
Chas. Neubert & Co., Baltimore, MD****
Mermaid Brand, Half Gallon, MD 39

Center left:
Chas. Neubert & Co., Baltimore, MD
Reverse of above can

Bottom left:
Chas. Neubert, Baltimore, MD***
Full Packt, Gallon, MD 39
Carlton and Mary Riggin Collection

Top right:
Chas. Neubert & Co., Baltimore, MD****
Mermaid Brand, Tall Quart and Pint
Ronald L. Newcomb Collection

Center right:
Neubert Bros., Baltimore, MD***
Pint, MD 32
Ronald L. Newcomb Collection

Bottom right:
J.D. Groves and Chas. Neubert Co.,
Baltimore, MD Can Lids

The Leib Packing Co., Baltimore, MD****
Sun Brand, Gallon, MD 27
Carlton and Mary Riggin Collection

The Leib Packing Co., Baltimore, MD****
Sun Brand, Tall Quart and Pint, MD 27
Ronald L. Newcomb Collection

J.J. Lansburgh & Co., Baltimore, MD****
Arrow Brand, Pint
Ronald L. Newcomb Collection

J.J. Lansburgh & Co., Baltimore, MD****
Arrow Brand, Tall Quart

Castle Oyster Co., Baltimore, MD***
Castle Brand, Gallon, MD 32
Ronald L. Newcomb Collection

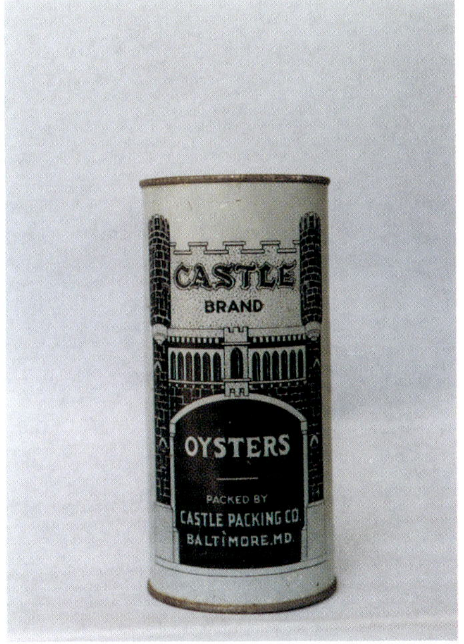

Castle Oyster Co., Baltimore, MD****
Castle Brand, Tall Quart

Jarrell & Rea, Baltimore, MD****
Silver Sea Brand, Pint
Carlton and Mary Riggin Collection

W.H. McGee & Co., Baltimore, MD****
Seal Brand, Half Gallon, MD 28
Carlton and Mary Riggin Collection

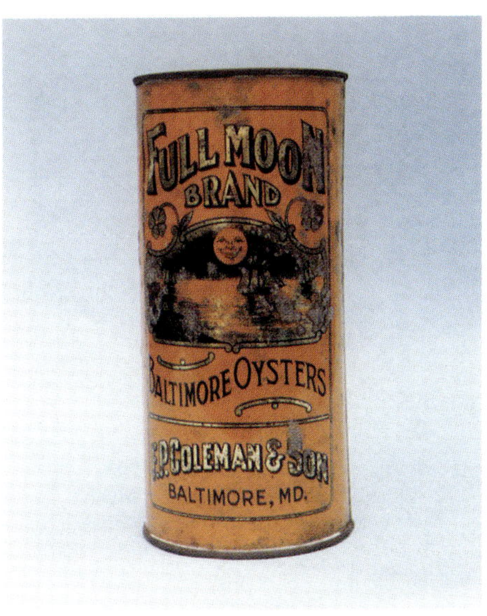

F.P. Coleman & Son, Baltimore, MD****
Full Moon Brand, Tall Quart
Mike and Eva Pinder Collection

Crescent Sea Food Co., Baltimore, MD**
Gallon, MD 26

W.H. McGee & Co., Baltimore, MD****
Seal Brand, Gallon
George and Betty Juergens Collection

F.P. Coleman & Son, Baltimore, MD****
Full Moon Brand, Gallon
Ronald L. Newcomb Collection

F.P. Coleman & Son, Baltimore, MD****
Full Moon Brand, Gallon
Bill and Steve Dorrell Collection

H.W. Ruth & Son, Grasonville, MD****
Bay Bridge Brand, Gallon, MD 175
Ronald L. Newcomb Collection

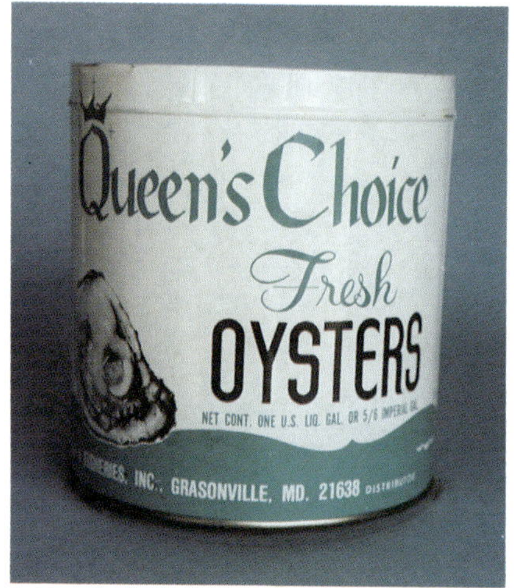

B & S Fisheries, Grasonville, MD**
Queen's Choice Brand, Gallon, MD 606
Carlton and Mary Riggin Collection

Islander Seafoods, Grasonville, MD****
Islander Brand, Gallon, MD 165
Courtesy of Harris Crab House

Fisherman's Seafood Market, Grasonville, MD**
Gallon, MD 163
Ronald L. Newcomb Collection

Calverts Shellfish Co., Grasonville, MD***
Calvert's Brand, Gallon, MD 536
Courtesy of Harris Crab House

A.C. Harris Co., Chester, MD****
Kent Island Brand, Half Gallon, MD 166
Mike and Eva Pinder Collection

E.H. Diggs & Son, Grasonville, MD****
Mary K Brand, Gallon, MD 160
Bill and Steve Dorrell Collection

Canabuci Seafoods, Grasonville, MD**
Blue Star Brand, Gallon MD 489
Carlton and Mary Riggin Collection

H.S. Thompson & Co., Grasonville, MD****
Stork Brand, 12 Ounces, MD 176
Bill and Steve Dorrell Collection

Second Stork Logo
H.S. Thompson & Co., MD 176
Bill and Steve Dorrell Collection

H.S. Thompson & Co., Grasonville, MD***
Stork Brand, Gallon, MD 176
Bill and Steve Dorrell Collection

Thomas & Thompson or H.S. Thompson & Co.***
Stork Brand, Gallon, MD 159 and MD 176
Grasonville, MD
Bill and Steve Dorrell Collection

W.H. Harris Seafood, Chester, MD**
Bay Shore Brand, Gallon, MD 177
Carlton and Mary Riggin Collection

Thomas & Thompson, Grasonville, MD****
Stork Brand, Pint
Mike and Eva Pinder Collection

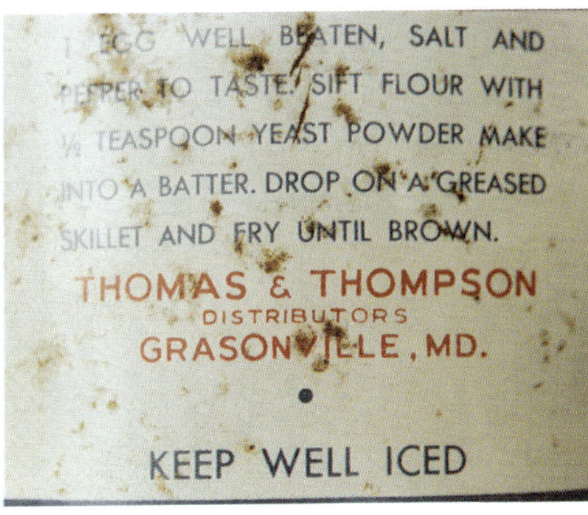

First Stork Logo
Thomas & Thompson, MD 159
Bill and Steve Dorrell Collection

Harrison Oyster Co., Tilghman, MD***
Harrison Brand, Gallon, MD 679
Gary and Sharon Campbell Collection

Harrison & Jarboe Seafood Co., St. Michaels, MD***
Miles River Brand, Gallon, MD 272
Ralph and Betty Tull Collection

Harrison & Jarboe Seafood Co., St. Michaels, MD***
Miles River Brand, Gallon, MD 273A
Ronald L. Newcomb Collection

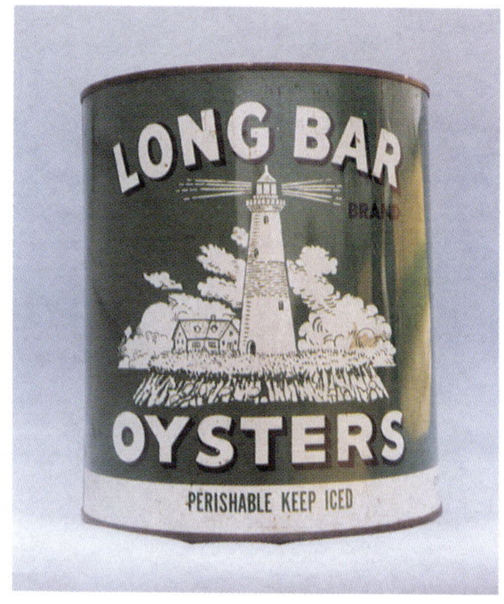

Oxford Packing Co., Oxford, MD****
Long Bar Brand, Gallon
Bill and Steve Dorrell Collection

Harrison & Jarboe Seafood Co., St. Michaels, MD***
Miles River Brand, Gallon, MD 272
Joe Sechrist Collection

Hanks Seafood Co., Easton, MD****
Cap-N Hanks Brand, Gallon, MD 80
Ronald L. Newcomb Collection

A.B. Harris & Co., Oxford, MD****
Tred Avon River Brand, Gallon
Gary and Sharon Campbell Collection

The Thos. E. Jones Co., Oxford, MD****
Oxford Brand, Gallon and Pint, MD 118

The Tilghman Packing Co., Tilghman, MD*****
Old Salt Brand, Gallon, MD 267
Ronald L. Newcomb Collection

Tilghman Packing Co., Tilghman, MD***
Tilghman Brand, Gallon, MD 267
Carlton and Mary Riggin Collection

Tilghman Packing Co., Tilghman, MD***
Tilghman Brand, Pint and Crab Meat Can
Courtesy of Nothing New Shop

Tilghman Pkg. Co., Tilghman, MD****
Tilghman Brand, Gallon

Tilghman Packing Co., Tilghman, MD***
Tilghman Brand, Gallon, MD 267
David C. Mitchell Collection

Tilghman Packing Co., Tilghman, MD****
Tilghman Brand, Half Gallon, MD 267
Carlton and Mary Riggin Collection

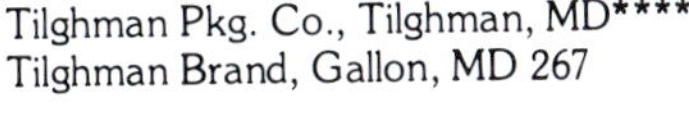
Tilghman Pkg. Co., Tilghman, MD****
Tilghman Brand, Gallon, MD 267

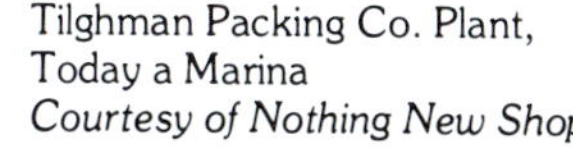
Tilghman Packing Co. Plant,
Today a Marina
Courtesy of Nothing New Shop

Plant of the Tilghman Packing Company situated on an Island made of oyster shells. Connected to mainland by long wooden bridge.

Secretary Oyster Co., Secretary, MD****
Philson Brand, Gallon, MD 171
Ronald L. Newcomb Collection

White & Nelson, Cambridge, MD***
Honga Brand, Gallon, MD 131
Carlton and Mary Riggin Collection

White & Nelson, Cambridge, MD***
Honga Brand, Small Indian Head Gallon, MD 112
Mike and Eva Pinder Collection

Philip J. Harrington & Son, Secretary, MD**
Harrington Brand, Gallon, MD 164
Carlton and Mary Riggin Collection

Philip J. Harrington & Son, Secretary, MD***
Harrington Brand, Gallon, MD 171

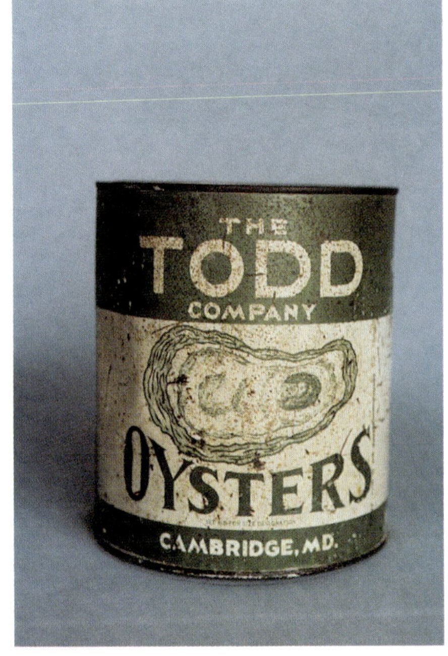

The Todd Co., Cambridge, MD****
Gallon, MD 128
Carlton and Mary Riggin Collection

Todd Seafood, Cambridge, MD***
Maryland's Finest, Gallon, MD 128
Carlton and Mary Riggin Collection

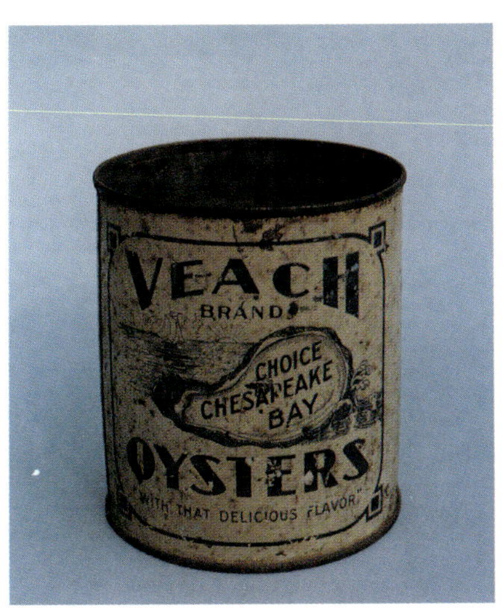

The J.M. Clayton Co., Cambridge, MD***
Veach Brand, Quart, MD 113
Courtesy of H.A.Fleckenstein, Jr.

Thos. E. Jones & Co., Cambridge, MD****
Jones Brand, Quart, MD 118
Ronald L. Newcomb Collection

I.L. Leonard & Co., Cambridge, MD***
Leonard's Brand, Pint, MD 119, Drinking Cup
Ronald L. Newcomb Collection

The J.M. Clayton Co., Cambridge, MD**
Epicure Brand, Gallon, MD 113
Carlton and Mary Riggin Collection

Mace, Woolford & Co., Cambridge, MD*****
Goose Creek Brand, Gallon
Ronald L. Newcomb Collection

I.L. Leonard & Co., Cambridge, MD***
Leonard's Brand, Pint, MD 119
Ronald L. Newcomb Collection

I.L. Leonard & Co., Cambridge, MD***
Leonard's Brand, Gallon, MD 119
Ronald L. Newcomb Collection

H.I. Brannock & Co., Cambridge, MD****
Brannock's Brand, Tall Quart
Mike and Eva Pinder Collection

Madison Seafood Co., Madison, MD**
B & J Brand, Gallon, MD 116
Carlton and Mary Riggin Collection

Meredith & Meredith, Wingate, MD**
Meredith's Brand, Gallon, MD 124
Carlton and Mary Riggin Collection

Bluepoints' Company, Cambridge, MD****
Bluepoints' Brand, Tall Quart, MD 275
Ronald L. Newcomb Collection

Chesapeake Shellfish Co., Sherwood, MD***
Harbor Cove Brand, Gallon, Md 273
Mike and Eva Pinder Collection

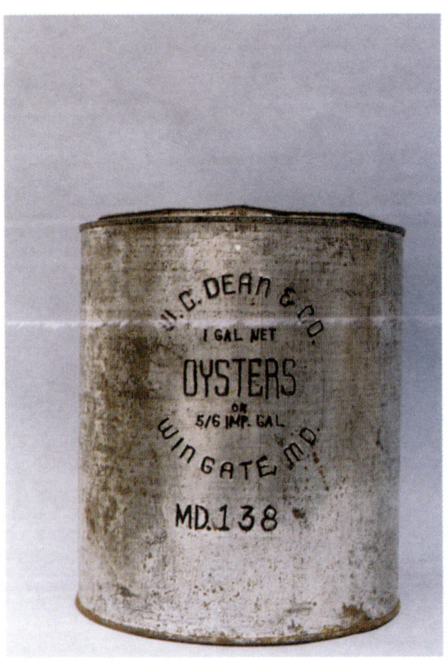

W, C. Dean Co. Wingate, MD**
Gallon, MD 138
Carlton & Mary Riggin Collection

The Salisbury Packing Co., Salisbury MD*****
Satisfaction Brand, Gallon, MD 290B
Courtesy of H. A. Fleckenstein, Jr.,

Ocean City Oyster Co., Ocean City, MD*****
Ocean Brand, Gallon, MD 291
Ronald L. Newcomb Collection

Roaring Point Oyster Co., Nanticoke, MD***
Sea Crest Brand, Gallon, MD 296
Ronald L. Newcomb Collection

Roaring Point Oyster Co., Nanticoke, MD***
Sea Crest Brand, Gallon, MD 296
Ronald L. Newcomb Collection

Nanticoke Seafood Co., Nanticoke, MD**
Maryland House Brand, Gallon,MD 136
Carlton and Mary Riggin Collection

Somerset Seafood Co., Deal Island, MD***
Gallon, MD 304
Carlton and Mary Riggin Collection

Nanticoke Seafood Co., Nanticoke, MD**
Maryland House Brand, Half Gallon, MD 228
Butch and Jackie Cheezum Collection

Bivalve Oyster Packing Co., Bivalve, MD**
Gallon, MD 290
Carlton and Mary Riggin Collection

H.B. Kennerly & Son, Nanticoke, MD**
Seacrest Brand, Gallon, VA 76

Bivalve Oyster Packing Co., Bivalve, MD**
B & L Brand, Gallon, MD 233
Carlton and Mary Riggin Collection

Anderson & Burton, Deal Island, MD**
Gallon, MD 250
Carlton and Mary Riggin Collection

Harold Bozman Seafood, Upper Fairmount, MD****
Manokin River Brand, Gallon, MD 610
David C. Mitchell Collection

Bivalve Oyster Packing Co., Bivalve, MD***
B & L Brand, Gallon, MD 289
Carlton and Mary Riggin Collection

Mt. Vernon Packing Co., Mt Vernon, MD**
G & E Brand, Gallon, MD 233
Carlton and Mary Riggin Collection

Zack Windsor & Co., Wenona, MD**
Gallon, MD 257
Carlton and Mary Riggin Collection

C.W. Howeth & Bro., Crisfield, MD****
H & B Brand, Pint, MD 193
Ronald L. Newcomb Collection

Metompkin Bay Oyster Co., Crisfield, MD**
Metopkin Brand, Gallon, MD 220
Carlton and Mary Riggin Collection

W.E. Riggin & Co., Crisfield, MD***
Pint, MD 205
Ronald L. Newcomb Collection

C.W. Howeth & Bro., Crisfield, MD***
H & B Brand, Gallon, MD 193
Carlton and Mary Riggin Collection

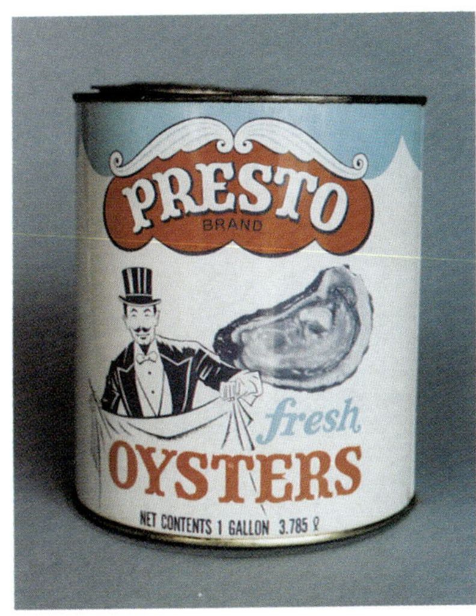

Franks Seafood, Crisfield, MD****
Presto Brand, Gallon, MD 68
Carlton and Mary Riggin Collection

Crisfield Supply Co., Crisfield, MD***
Pint, MD 240

Metompkin Bay Oyster Co., Crisfield, MD**
Metompkin Brand, Gallon, MD 220
Carlton and Mary Riggin Collection

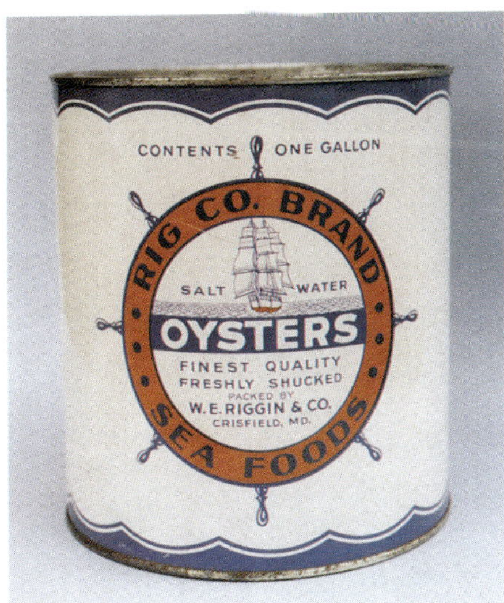

W.E. Riggin & Co., Crisfield, MD****
Rig Co. Brand, Gallon, MD 205
Ronald L. Newcomb Collection

Packer Unknown, Crisfield, MD***
Quality Brand, Gallon, MD 208
Ronald L. Newcomb Collection

George A. Christy & Son, Crisfield, MD****
Christy's Brand, Gallon, MD 223
Randy Schreck Collection

Milbourne Oyster Co., Crisfield, MD**
Gallon, MD 364
Carlton and Mary Riggin Collection

L.R. Carson, Crisfield, MD***
Quality Brand, Gallon, MD 243
Carlton and Mary Riggin Collection

George A. Christy & Son, Crisfield, MD***
Christy's Brand, Gallon, MD 223
Gary and Sharon Campbell Collection

Milbourne Oyster Co., Crisfield, MD***
Milbourne's Brand, Gallon, MD 203
Carlton and Mary Riggin Collection

L.R. Carson, Crisfield, MD**
Gallon, MD 243
Carlton and Mary Riggin Collection

Geo. A. Christy & Sons, Crisfield, MD***
C Brand, Pint
Bill and Steve Dorrell Collection

Milbourne Oyster Co., Crisfield, MD***
MOCO Brand, Pint, MD 203
Ronald L. Newcomb Collection

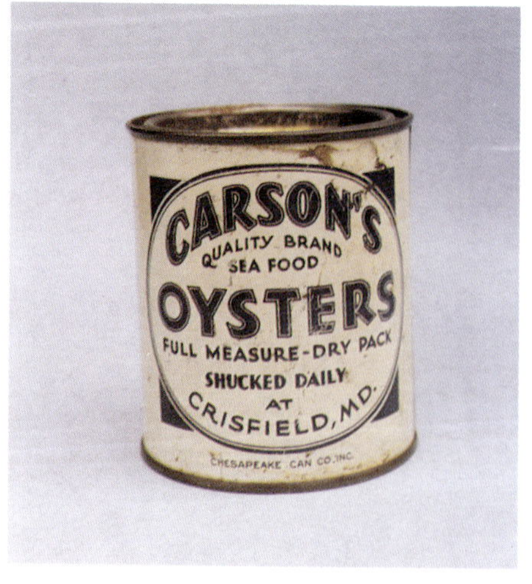

L.R. Carson, Crisfield, MD***
Quality Brand, Quart, MD 243
Ronald L. Newcomb Collection

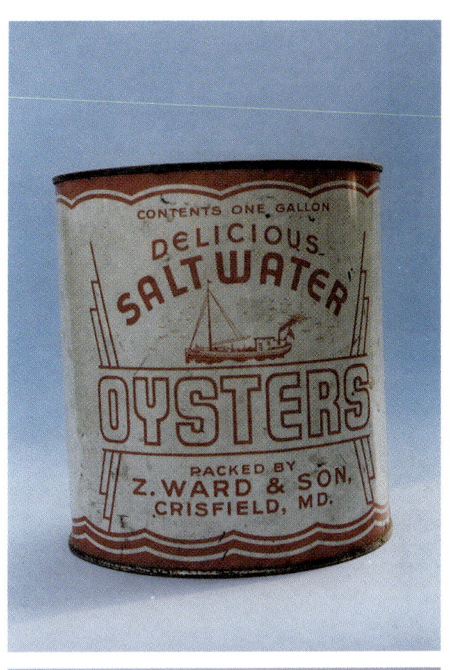

Z. Ward & Son, Crisfield, MD****
Gallon, MD 217
Carlton and Mary Riggin Collection

Packer Unknown, Crisfield, MD****
Possible Ward Oyster Co.
Star Brand, Pint, MD 198

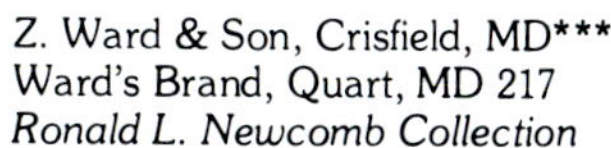
Z. Ward & Son, Crisfield, MD***
Ward's Brand, Quart, MD 217
Ronald L. Newcomb Collection

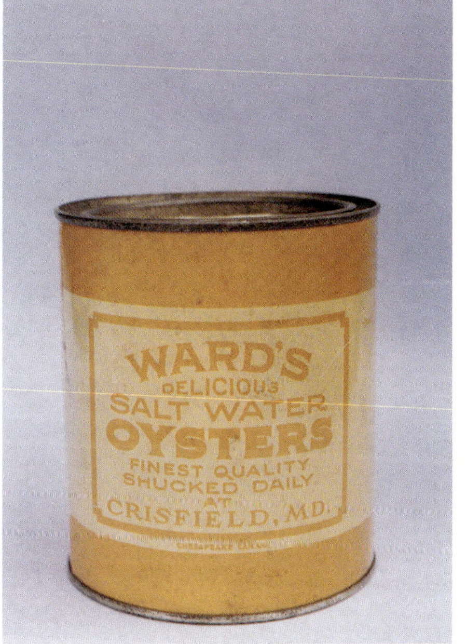

Packer Unknown, Crisfield, MD****
M & V Brand, Pint, MD 199

Reliable Seafood Co., Crisfield, MD***
Reliable Brand, Gallon, MD 239
A Brand of Helsby Seafood Co.
Ronald L. Newcomb Collection

Helsby Seafood Co., Crisfield, MD**
Helsby Brand, Gallon, MD 239

John T. Handy Co., Crisfield, MD****
Gallon, MD 224
Ronald L. Newcomb Collection

Back of Can Above

John T. Handy Co., Crisfield, MD***
Handy's Brand, Pint, MD 224
Ronald L. Newcomb Collection

John T. Handy Co., Crisfield, MD**
Handy Brand, 12 Ounces, MD 224

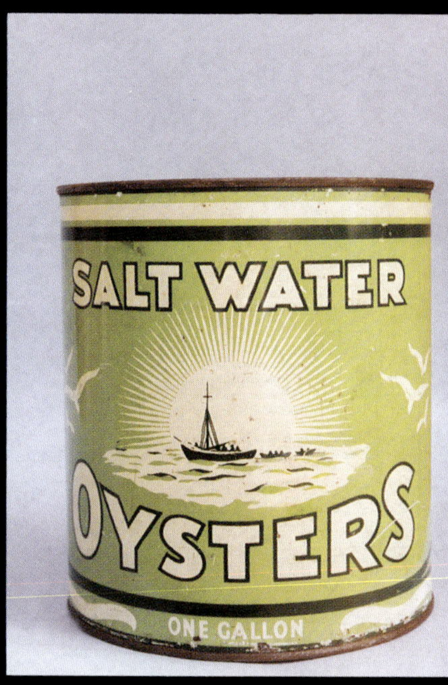

John T. Handy Co., Crisfield, MD****
Gallon, MD 224

John T. Handy Co., Crisfield, MD***
Pint, Drinking Cup, MD 224
Note Slogans "The Essence of Goodness" and "The
Key to that Paradise Called Appetite"
Copywrite Given in 1922 to the Goodwin Oyster Co
Columbus, OH
Bill and Steve Dorrell Collection

Carol Dryden & Co., Crisfield, MD****
Dryden's Quality Brand, Tall Quart, MD 196

Carol Dryden & Co., Crisfield, MD****
Pride of the Chesapeake Brand, Tall Quart and Pint, MD 196
Ronald L. Newcomb Collection

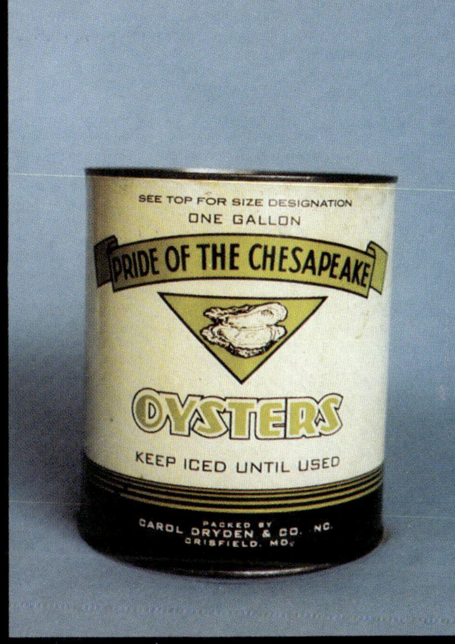

Carol Dryden & Co., Crisfield, MD***
Pride of the Chesapeake Brand, Gallon, MD 196
Carlton and Mary Riggin Collection

Carol Dryden & Co., Crisfield, MD***
Pride of the Chesapeake, Pint, MD 196

Carol Dryden & Co., Crisfield, MD****
Dryden's Quality Brand, Pint, MD 196
Carlton and Mary Riggin Collection

N.R. Coulbourn, Crisfield, MD****
Deep-Rock Brand, Gallon and Quart
Carlton and Mary Riggin Collection

N.R. Coulbourn, Crisfield, MD****
Deep Rock Brand, Pint
Carlton and Mary Riggin Collection

N.R. Coulbourn, Crisfield, MD***
Deep Rock Brand, Gallon
Carlton and Mary Riggin Collection

Crisfield Packing Co., Crisfield, MD**
Gallon, MD 199
Ralph and Betty Tull Collection

V.L. Evans & Co., Crisfield, MD***
Evans Brand, Gallon, MD 235

N.R. Coulbourn, Crisfield, MD****
Deep Rock Brand, Tall Quart
Butch and Jackie Cheezum Collection

E.R. Dize & Co., Crisfield, MD***
MAR-VA Brand, Pint, MD 206
Bill and Steve Dorrell Collection

I.W. Lawson & Co., Crisfield, MD***
Gallon, MD 226
Ronald L. Newcomb Collection

Lawrence Tyler & Co., Crisfield, MD****
Tyler's Brand, Gallon, MD 236
Courtesy of Harris Crab House

Crockett & Forbush, Crisfield, MD*****
Black Pearl Brand, Gallon, MD 220
Bill and Steve Dorrell Collection

C.A. Loockerman, Crisfield, MD****
Originalpac Brand, Gallon, MD 191
Ronald L. Newcomb Collection

Lawson Oyster Co., Crisfield, MD***
Lawson Brand, Gallon, MD 226
Ronald L. Newcomb Collection

Hickman & Sterling, Crisfield, MD**
Gallon, MD 202
Carlton and Mary Riggin Collection

Sterling and Somers, Crisfield, MD***
S & S Brand, Pint, MD 216
Ronald L. Newcomb Collection

W.E. Ward Oyster Co., Crisfield, MD***
Ward Brand, Pint, MD 241
Bill and Steve Dorrell Collection

Booth Fisheries Co., Crisfield, MD***
Booth Brand, Gallon
David C. Mitchell Collection

Ray's Seafood, Crisfield, MD**
Gallon, MD 570
Carlton and Mary Riggin Collection

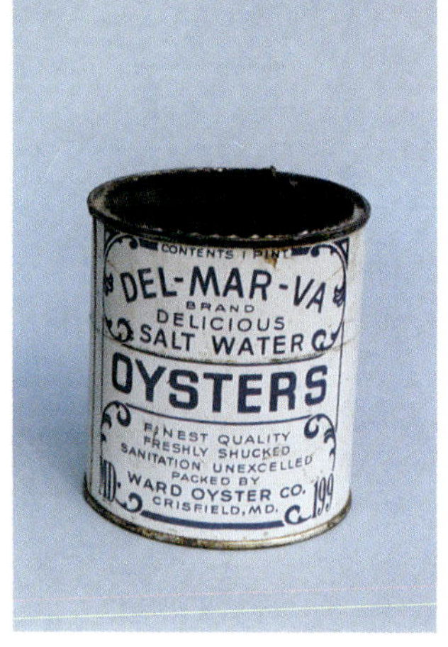

Ward Oyster Co., Crisfield, MD***
Del-Mar-Va Brand, Pint, MD 199
Courtesy of H.A. Fleckenstein, Jr.

J. Lloyd Sterling & Co., Crisfield, MD***
Gallon, MD 215 with pair of J. Lloyd Sterling Miniature Baldpate Decoys
Carlton and Mary Riggin Collection

Seafood Co-Op Inc., Shady Side, MD*
Gallon, MD 77 Successor to Leatherbury Bros.
Carlton and Mary Riggin Collection

Leatherbury Bros., Shady Side, MD**
Black Swan Brand, Gallon, MD 77

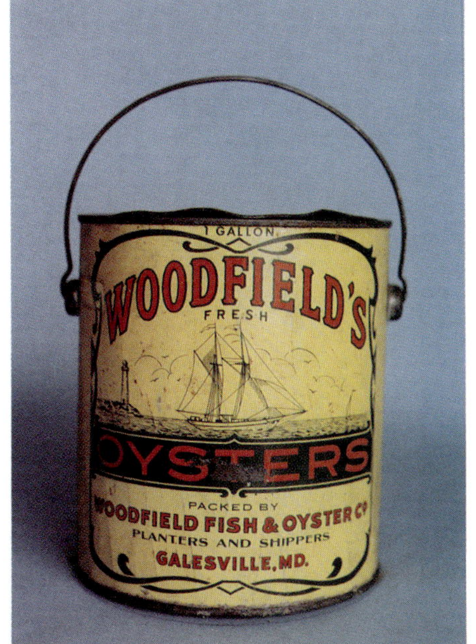

Woodfield Fish & Oyster Co., Galesville, MD****
Chesapeake Brand, Gallon
Carlton and Mary Riggin Collection

Denton & Rogers, Broomes Island, MD****
Calvert Brand, Gallon, MD 96
Joe Sechrist Collection

Woodfield Fish & Oyster Co., Galesville, MD****
Woodfield's Brand, Gallon
Carlton and Mary Riggin Collection

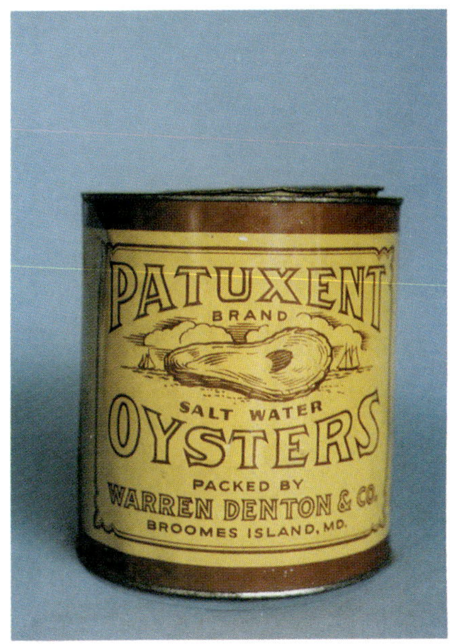

Warren Denton & Co., Broomes Island, MD**
Patuxent Brand, Gallon, MD 96
Carlton and Mary Riggin Collection

Leonard Copsey, Oraville, MD***
Patuxent River Brand, Gallon, MD 191
Carlton and Mary Riggin Collection

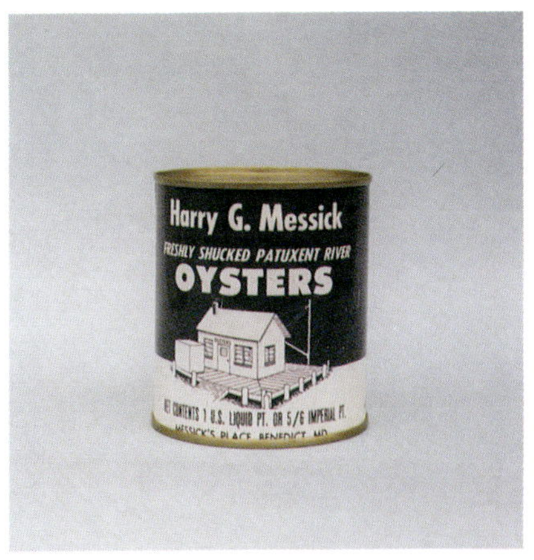

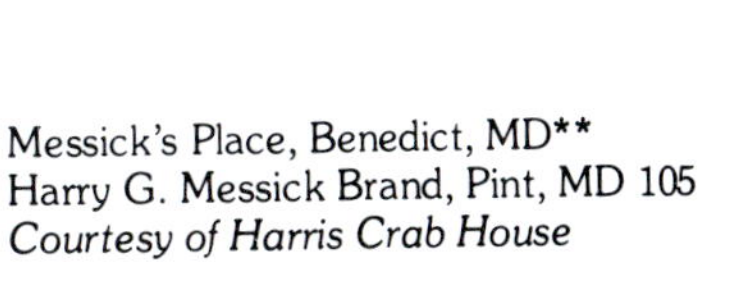

Messick's Place, Benedict, MD**
Harry G. Messick Brand, Pint, MD 105
Courtesy of Harris Crab House

H.M. Woodburn & Son, Solomons, MD****
A M W Brand, Gallon, MD 102
Ronald L. Newcomb Collection

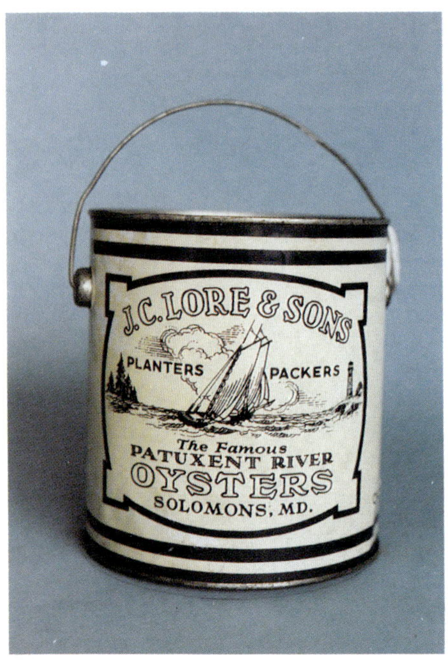

J.C. Lore & Sons, Solomons, MD****
Gallon
Carlton and Mary Riggin Collection

McNasby Oyster Co., Annapolis, MD****
Pearl Brand, Gallon
Courtesy of Dick Larrimore

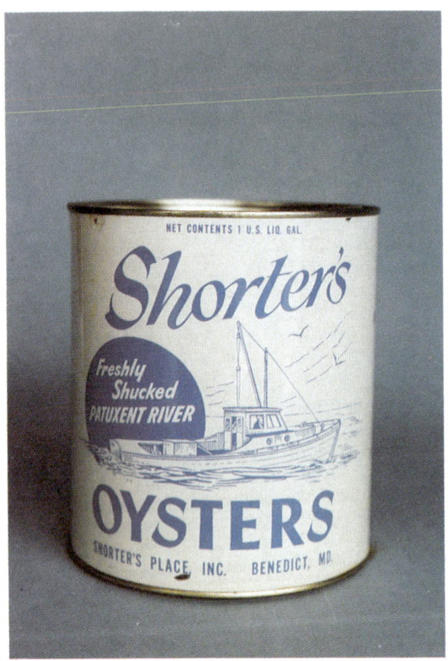

Shorter's Place, Inc., Benedict, MD***
Shorter's Brand, Gallon, MD 103
Carlton and Mary Riggin Collection

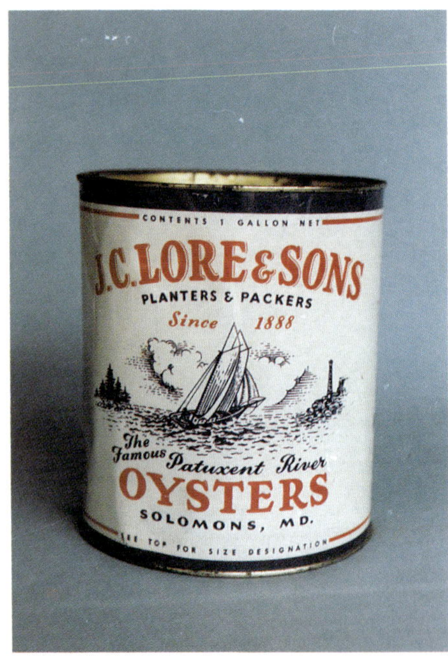

J.C. Lore & Sons, Solomons, MD***
Gallon, MD 101
Carlton and Mary Riggin Collection

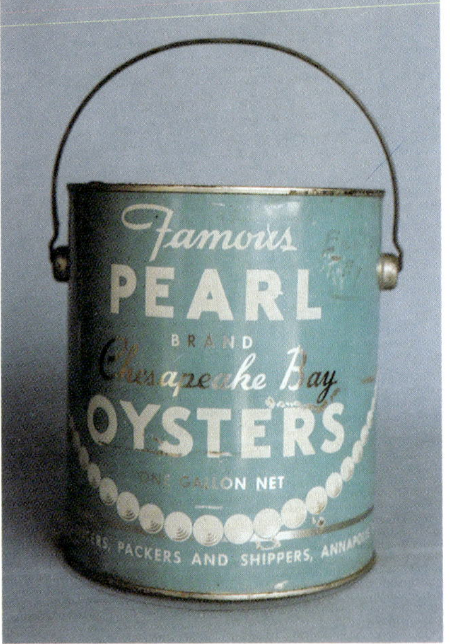

McNasby Oyster Co., Annapolis, MD****
Pearl Brand, Gallon
Carlton and Mary Riggin Collection

J.C. Lore & Sons, Solomons, MD***
Pint, MD 101

Charles E. Davis, Wynne, MD***
Potomac View Brand, Gallon, MD 183
Ronald L. Newcomb Collection

VIRGINIA

Wm. C. Bunting, Chincoteague, VA***
Tom's Cove Brand, Gallon, VA 437
Carlton and Mary Riggin Collection

McCready Bros., Chincoteague, VA***
McCready Brand, Gallon, VA 198
Carlton and Mary Riggin Collection

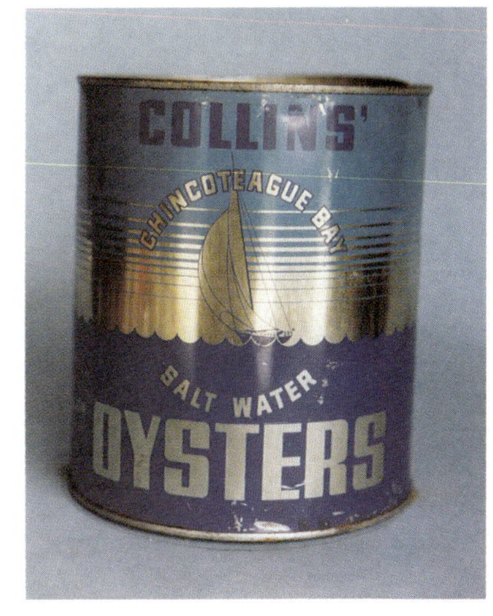

Nelson P. Collins, Chincoteague, VA***
Collins' Brand, Gallon, VA 25
Carlton and Mary Riggin Collection

Nelson P. Collins, Chincoteague, VA***
Chincoteague Bay Brand, Gallon, VA 25
Carlton and Mary Riggin Collection

Ralph E. Watson Oyster Co., Chincoteague, VA***
Tom's Cove Brand, Gallon, VA 283
Carlton and Mary Riggin Collection

Wm. C. Bunting Oyster Co., Chincoteague, VA***
Tom's Cove, Gallon, VA 437
Carlton and Mary Riggin Collection

Wm. C. Bunting, Chintoteague, VA***
Tom's Cove Brand, Gallon, VA 437
Carlton and Mary Riggin Collection

Chincoteague Oysters, Chincoteague, VA**
Gallon and Half Gallon, VA 312 & VA 235

Eslie Justice, Chincoteague, VA**
Gallon, VA 2
Carlton and Mary Riggin Collection

Shreves Bros., Chincoteague, VA***
Gallon, VA 312

Burton's Seafood, Chincoteague, VA***
Gallon, VA 52
Carlton and Mary Riggin Collection

W.E.Jones Seafood, Chincoteague, VA***
Gallon, VA 511
Gary and Sharon Campbell Collection

Earl B. Hill, Chincoteague, VA**
Gallon, VA 351
Carlton and Mary Riggin Collection

Savage & Mears, Chincoteague, VA**
S & M Brand, Gallon, VA 235
Carlton and Mary Riggin Collection

Jones Brothers Seafood, Chincoteague, VA***
Jones Brothers Brand, Gallon, VA 136
Carlton and Mary Riggin Collection

Thornton Bros. Seafood, Chincoteague, VA**
Gallon, VA 401
Carlton and Mary Riggin Collection

Savage & Mears, Chincoteague, VA**
Gallon, VA 151
Carlton and Mary Riggin Collection

Conner & McGee, Chincoteague, VA***
Gallon, VA 327
Carlton and Mary Riggin Collection

Seaside Oyster Co., Chincoteague, VA***
Sal-T-Sea Brand, Gallon, VA 20
Gary and Sharon Campbell Collection

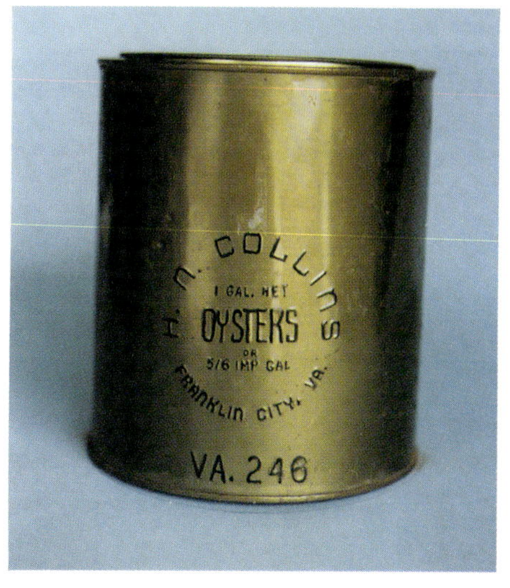

H.N. Collins, Franklin City, VA***
(Franklin City no Longer Exists)
Gallon, VA 246
Carlton and Mary Riggin Collection

Reginald Stubbs Seafood Co., Chincoteague, VA**
Chincoteague Island Brand, Gallon, VA 223

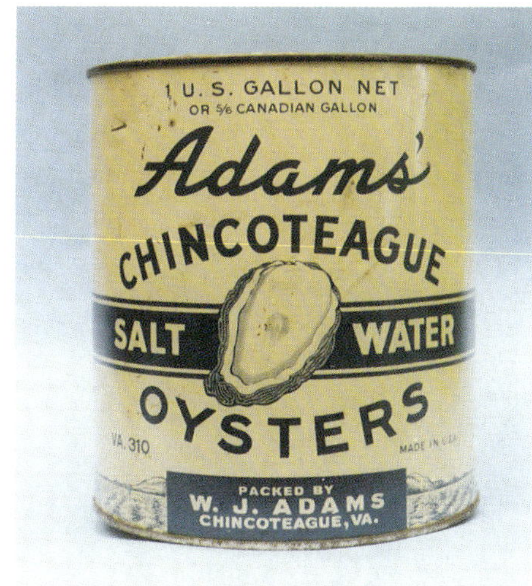

W.J. Adams, Chincoteague, VA***
Adams' Brand, Gallon, VA 310
Gary and Sharon Campbell Collection

Grady Rhodes Seafood, Saxis, VA****
Island Rock Brand, Gallon VA 48
Keeford & Martha Linton Collection

Tom's Cove Oyster Co., Chincoteague, VA***
Tom's Cove Brand, Gallon, VA 585
Ronald L. Newcomb Collection

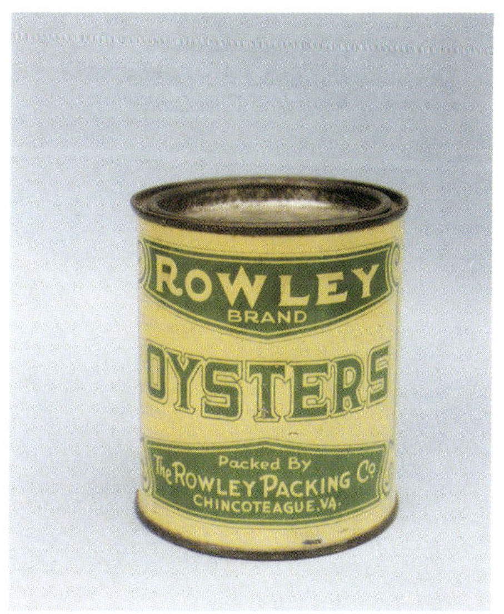

The Rowley Packing Co., Chincoteague, VA***
Rowley Brand, Pint, VA 17
Ronald L. Newcomb Collection

Milton Drewer, Saxis, VA***
Gallon, VA 600

H.M. Terry Co., Willis Wharf, VA***
Sewansecott Brand, Gallon, VA 47
Carlton and Mary Riggin Collection

J.C. Walker Bros., Exmore, VA***
Walker's Seaside Brand, Gallon, VA 172
Carlton and Mary Riggin Collection

J.C. Walker Bros., Exmore, VA***
Walker's Seaside Brand, Pint
Carlton and Mary Riggin Collection

J.C. Walker Bros., Exmore, VA***
Virginia Seaside Brand, Pint, VA 172
Bill and Steve Dorrell Collection

J.C. Walker Bros., Exmore, VA***
Walker's Seaside Brand, Gallon, VA 172
Carlton and Mary Riggin Collection

J. C. Walker, Willis Wharf, VA****
Virginia Seaside Brand, Gallon, VA 172
Carlton and Mary Riggin Collection

Ballard Bros. Fish Co., Willis Wharf, VA***
Egg Island Brand, Half Gallon, VA 171
Carlton and Mary Riggin Collection

Ballard Bros. Fish Co., Willis Wharf, VA***
Gallon, VA 171
Carlton and Mary Riggin Collection

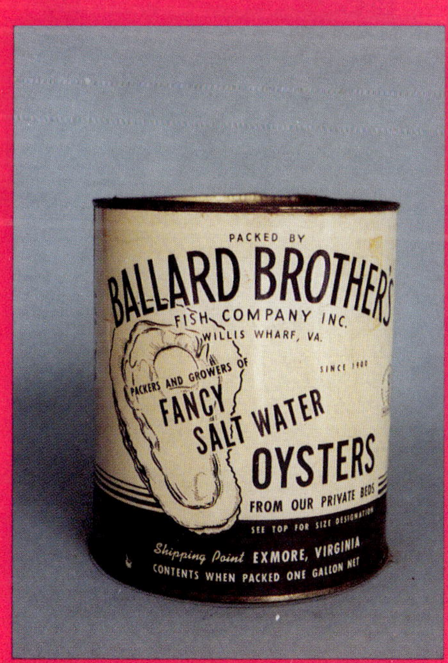

Ballard Brothers Fish Co., Willis Wharf, VA***
Gallon, VA 171
Carlton and Mary Riggin Collection

Ballard Bros. Fish Co., Willis Wharf, VA***
Egg Island Brand, Gallon, VA 171
Carlton and Mary Riggin Collection

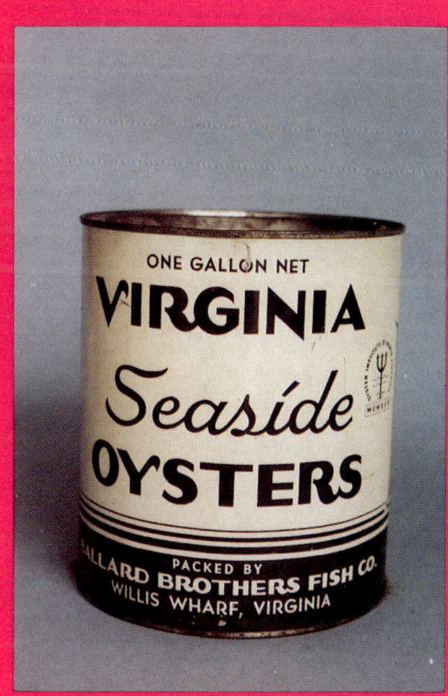

Ballard Brothers Fish Co., Willis Wharf, VA***
Virginia Seaside Brand, Gallon, VA 171
Carlton and Mary Riggin Collection

West Oyster Co., Oyster, VA***
West's Brand, Gallon, VA 11
Carlton and Mary Riggin Collection

Bonniville & Edgerton, Cape Charles, VA*****
Gallon
Carltonand Mary Riggin Collection

E.J. Steelman's, Cheriton, VA***
Cherrystone Brand, Gallon, VA 371
Bill and Steve Dorrell Collection

Ralph A. Clark & Son, Bridgetown, VA**
Gallon, VA 259
Carlton and Mary Riggin Collection

N.R. Steelman, Oyster, VA**
Gallon, VA 164
Carlton and Mary Riggin Collection

H. Allen Smith, Cheriton, VA**
Gallon, VA 585
Carlton and Mary Riggin Collection

D.N. Wisherd Sons, Oyster, VA****
Wisherd's Brand, Pint
Mike and Eva Pinder Collection

Ballard Fish & Oyster Co., Norfolk, VA***
Egg Island Brand, Gallon, VA 207
Carlton and Mary Riggin Collection

Ballard Fish & Oyster Co., Norfolk, VA***
Egg Island Brand, Gallon, VA 207
Carlton and Mary Riggin Collection

W.J. Crosby & Co., Norfolk, VA***
Sea Garden Brand, Pint, VA 5
Ronald L. Newcomb Collection

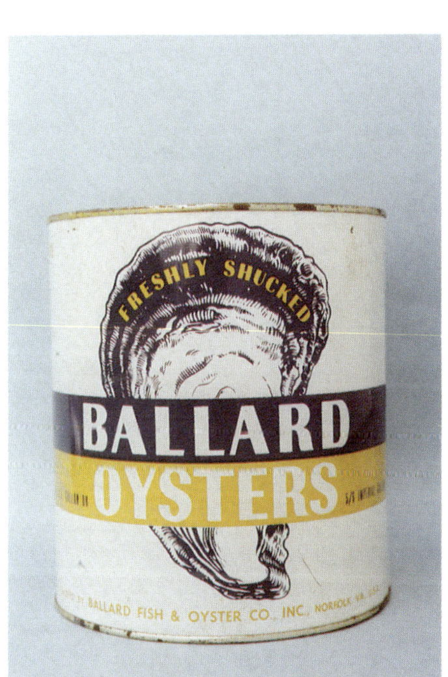

Ballard Fish & Oyster Co., Norfolk, VA***
Ballard Oysters Brand, Gallon, VA 207
Carlton and Mary Riggin Collection

Ballard-Type Lower Delaware Bay,****
Egg Island Brand, Gallon, NJ 5
Carlton and Mary Riggin Collection

W.J. Crosby & Co., Norfolk, VA***
Virginia Brand, Pint
Mike and Eva Pinder Collection

J.H. Miles & Company, Norfolk, VA****
Sea-Kist Brand, Gallon, VA 214
Ronald L. Newcomb Collection

J.H. Miles & Company, Norfolk, VA****
Ocean View Brand, Gallon, VA 214
Carlton and Mary Riggin Collection

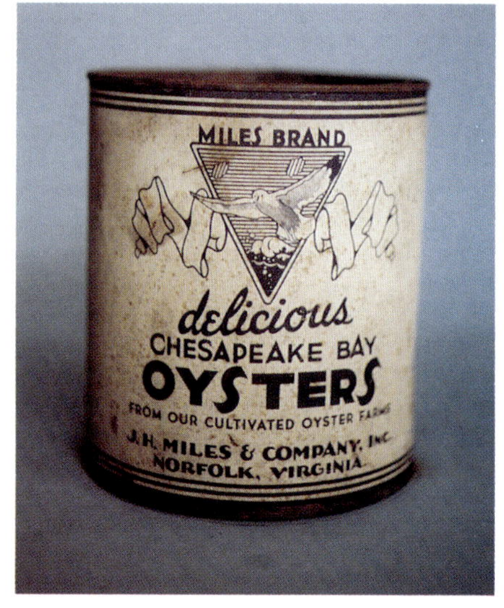

J.H. Miles & Company, Norfolk, VA***
Miles Brand, Gallon, VA 214
Carlton and Mary Riggin Collection

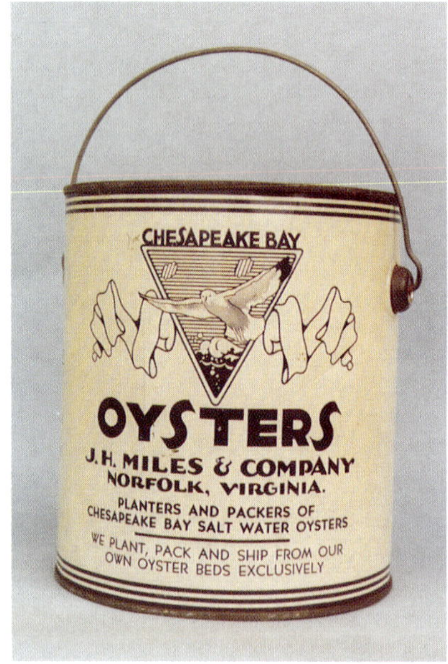

J.H. Miles & Company, Norfolk, VA****
Chesapeake Bay Brand, Gallon
Butch and Jackie Cheezum Collection

J.H. Miles & Company, Norfolk, VA***
Ocean View Brand, Gallon, VA 214
Carlton and Mary Riggin Collection

J.H. Miles & Company, Norfolk, VA***
Miles Brand, Gallon, VA 214
Carlton and Mary Riggin Collection

Lid for Can Above

J.H. Miles & Company, Norfolk, VA**
Miles Brand, Gallon, VA 214
Carlton and Mary Riggin Collection

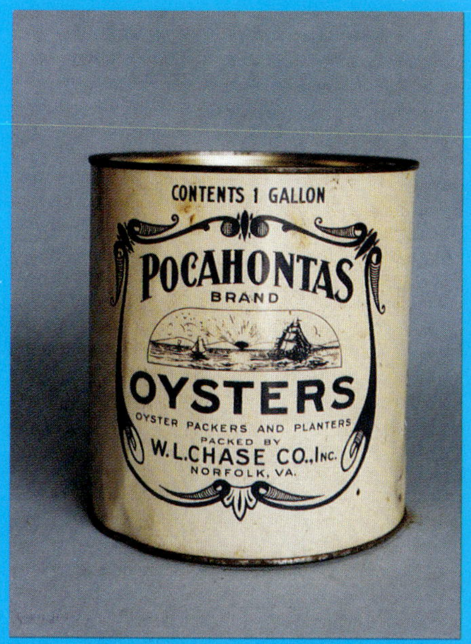

W.L. Chase Co., Norfolk, VA***
Pocahontas Brand, Gallon, VA 209
Carlton and Mary Riggin Collection

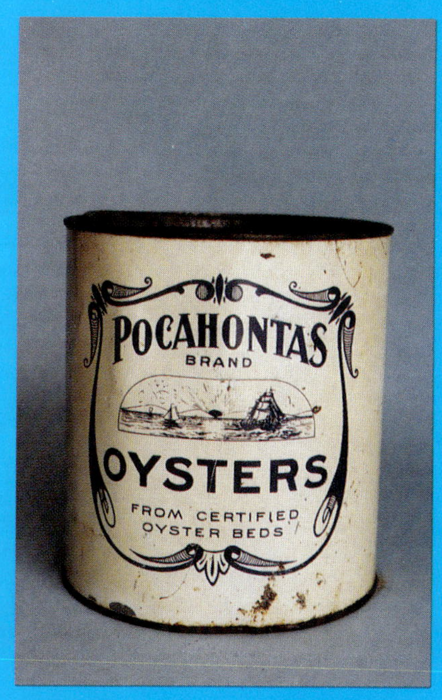

W.L. Chase Co., Norfolk, VA***
Pocahontas Brand, Gallon, VA 209
Carlton and Mary Riggin Collection

W.L. Chase Co., Norfolk, VA**
Chase Brand, Gallon, VA 214
Courtesy of Harris Crab House

Packer Unknown****
NorVa Brand, Tall Quart

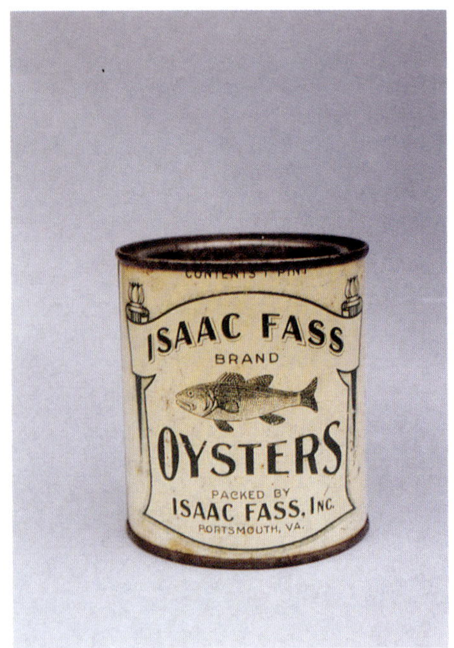

Isaac Fass, Portsmouth, VA***
Isaac Fass Brand, Pint, VA 221
Ronald L. Newcomb Collection

Hogg's Oyster Co., Gloucester Point, VA**
Hogg's Brand, Gallon, VA 89

Fast Bros., Hampton, VA***
Virginia Capes Brand, Gallon, VA 291
Carlton and Mary Riggin Collection

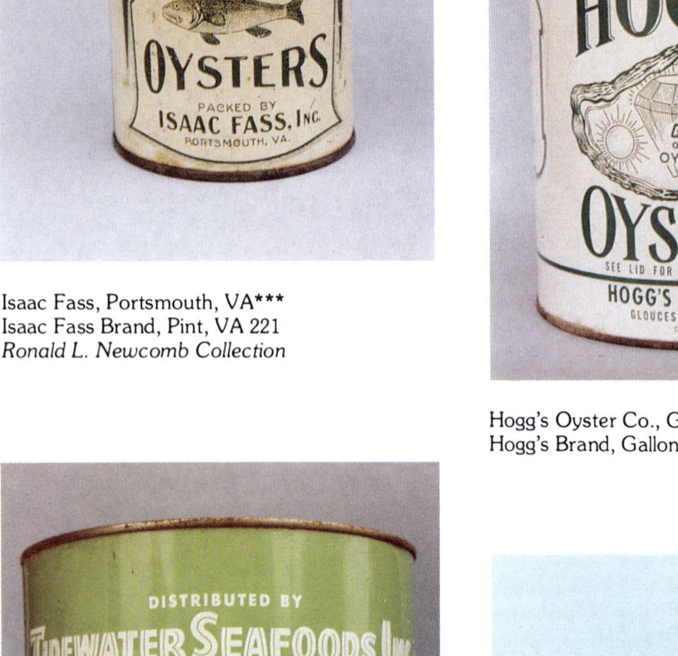

Tidewater Seafoods, Newport News, VA***
Gallon, VA 570
Ronald L. Newcomb Collection

F.C. Hogg, Gloucester Point, VA**
Gallon, VA 89

Virginia Seafood Exchange, Newport News, VA***
Virginia Sea Brand, Gallon, VA 87
Ronald L. Newcomb Collection

J.S. Darling & Son, Hampton VA***
Darling's Brand, Gallon, VA 58
Carlton and Mary Riggin Collection

M.F. Quinn, Hampton, VA*****
Plum Tree Island Brand, Gallon, VA 347
Bill and Steve Dorrell Collection

Cook's Seafood, Hampton, VA***
Back River Seafood Brand, Gallon, VA 350
Carlton and Mary Riggin Collection

G. T. Elliott, Inc., Hampton, VA***
Elliott's Brand, Gallon, VA 59
Courtesy of Black Swan Antiques

Curley Packing Co., Colonial Beach, VA***
Curley's Brand, Gallon, VA 167
Bill and Steve Dorrell Collection

Herbert Wilkerson & Son, Colonial Beach, VA***
Wilkerson's Brand, Gallon, VA 139
David C. Mitchell Collection

Battery Park Fish & Oyster Co., Battery Park, VA****
Oasis Brand, Gallon, VA 89
Mike and Eva Pinder Collection

W.H. & G.B. Sparrer, Hampton VA***
Sparrer Brand, Gallon, VA 607
Carlton and Mary Riggin Collection

Battery Park Fish & Oyster Co., Battery Park, VA***
Oceanspra Brand, Gallon, VA 89
Carlton and Mary Riggin Collection

Battery Park Fish & Oyster Co., Battery Park, VA***
Oceanspra Brand, Gallon, VA 89
Carlton and Mary Riggin Collection

W.H. & G.B. Sparrer, Hampton VA****
Sparrer Brand, Gallon, VA 607
Carlton and Mary Riggin Collection

York River Seafood Co., Seaford, VA***
Bay Brand, Gallon, VA 256
Carlton and Mary Riggin Collection

Cuthbert & Hughes, Bohannon, VA***
Old Virginia Brand, Gallon, VA 66
Ronald L. Newcomb Collection

Cook Seafood Co., York, VA**
York Brand, Gallon, VA 394
Carlton and Mary Riggin Collection

York River Oyster Corp., Gloucester, VA****
Williamsburg Brand, Gallon

Ware River Seafood, Schley, VA***
Ware River Brand, Gallon, VA 672
Ronald L. Newbomb Collection

York River Seafood Co., Hayes Store, VA**
Gallon, VA 82
Carlton and Mary Riggin Collection

J.W. Ferguson Seafood Co., Remlick, VA***
Rappahannock River Brand, Gallon, VA 17
Carlton and Mary Riggin Collection

J.W. Hurley & Son, Urbanna, VA****
Gallon, VA 139
Mike and Eva Pinder Collection

J.W. Hogge Seafood Co., Hayes, VA**
River's Delight Brand, 12 Ounces, VA 169

Hale Seafoods, Morattico, VA***
Big H Brand, Gallon, VA 107
Butch and Jackie Cheezum Collection

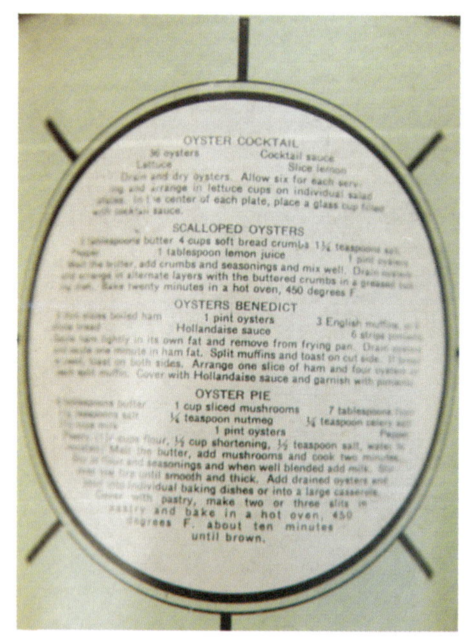

Back of Above Can Showing Recipes

Lancaster Seafoods, Morattico, Va***
Gallon, VA 107
Carlton and Mary Riggin Collection

C.T. Slaughter, Morattico, VA****
Big S Brand, Gallon, VA 107
Mike and Eva Pinder Collection

Piney Island Seafood, Morattico, VA***
SEAsa-weHAK Brand, Gallon, VA 168
Ralph and Betty Tull Collection

Lancaster Seafoods, Morattico, VA***
Gallon, VA 107
Ronald L. Newcomb Collection

C.T. Slaughter, Morattico, VA***
Gallon, VA 107
Carlton and Mary Riggin Collection

RCV Seafood, Morattico, VA***
RCV Brand, Gallon, VA 107
Bill and Steve Dorrell Collection

C.P. Saunders & Son, Millenbeck, VA**
Capt. Boyd's Brand, Gallon, VA 142
Courtesy of Harris Crab House

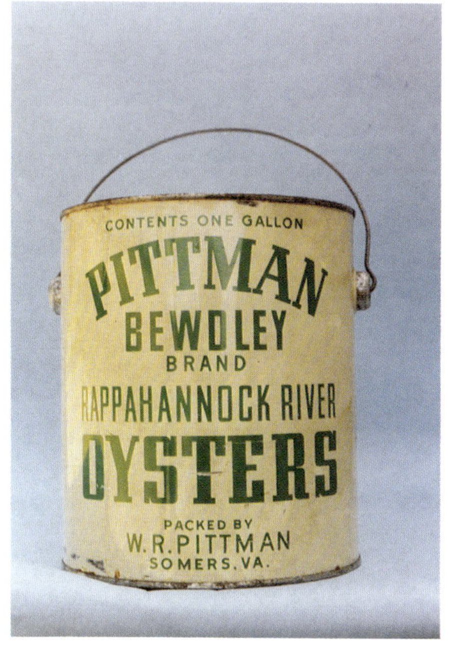

W.R. Pittman, Somers, VA****
Pittman Bewdley Brand, Gallon, VA 10
Bill and Steve Dorrell Collection

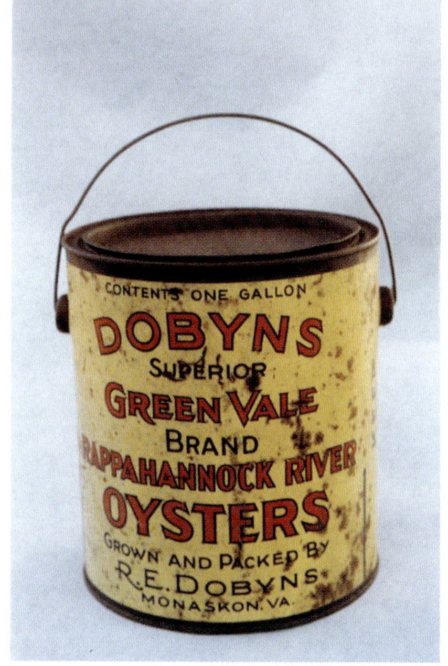

R.E. Dobyns, Monaskon, VA****
Green Vale Brand, Gallon, VA 105
Carlton and Mary Riggin Collection

Bevans Oyster Co., Kinsale, VA**
Bevans Brand, Pint, VA 303
Mike and Eva Pinder Collection

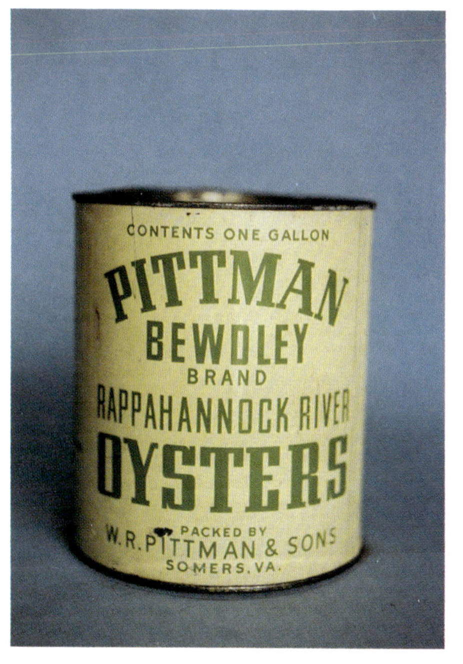

W.R. Pittman & Sons, Somers, VA***
Pittman Bewdley Brand, Gallon, VA 10
Carlton and Mary Riggin Collection

E.J. Conrad & Sons Seafood, Lancaster, VA***
Green Vale Brand, Gallon, VA 318
Carlton and Mary Riggin Collection

T.A. Treakle & Son, Palmer, VA**
Pint, VA 113

Adam's Packing Corp., Grimstead, VA***
APCO Brand, Pint, VA 515
Bill and Steve Dorrell Collection

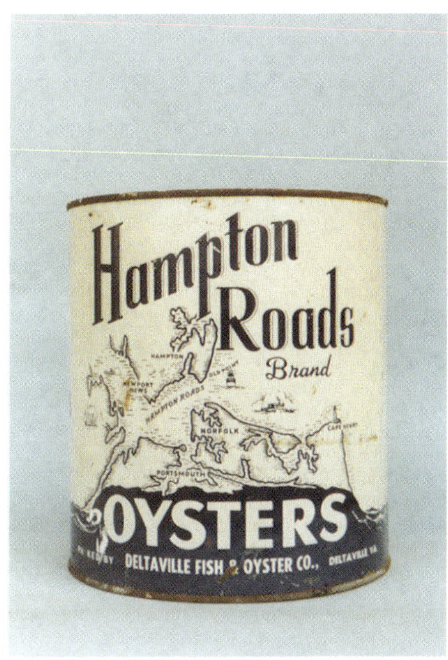

Deltaville Fish & Oyster Co., Deltaville, VA****
Hampton Roads Brand, Gallon, VA 146
Butch and Jackie Cheezum Collection

Haywood Oyster Co., Perrin, VA***
Ocean Breeze Brand, Gallon, VA 123
Carlton and Mary Riggin Collection

J. Newton Foster, Grimstead, VA**
Gallon, VA 515
Carlton and Mary Riggin Collection

R.E. Naumann Sea Food, Senora, VA***
Naumann Brand, Gallon, VA 73
Courtesy of Harris Crab House

B.G. Smith & Son, Sharps, VA***
Perch Creek Brand, Gallon, VA 230
Carlton and Mary Riggin Collection

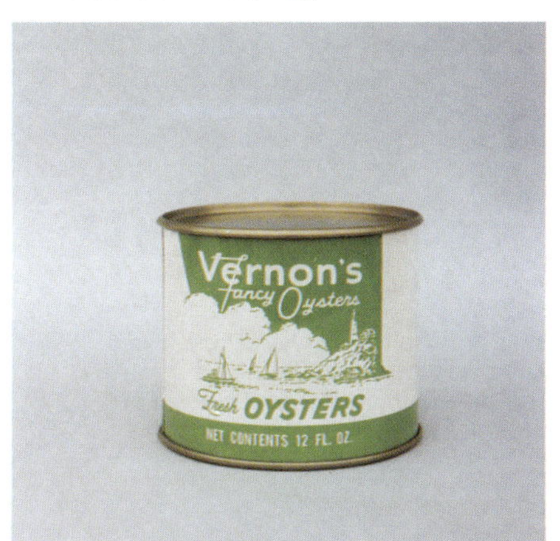

V.A. Haywood Seafoods, Perrin, VA**
Vernon's Brand, 12 Ounces, VA 227
Courtesy of Harris Crab House

Waterview Packing Co., Waterview, VA***
Tasty Brand, Gallon, VA 536
Carlton and Mary Riggin Collection

J.W. Ferguson Seafood Co., Remlick, VA***
Choice of Chesapeake Bay Brand, Half Gallon, VA 17
Butch and Jackie Cheezum Collection

Crockett Seafood Inc., Irvington, VA***
Davy Crockett Brand, Gallon, VA 445
Carlton and Mary Riggin Collection

Earl Cockrell, Burgess, VA**
Gallon, VA 183
Carlton and Mary Riggin Collection

Virginia Seafoods, Irvington, VA****
King Carter Brand, Gallon, VA 84
Ronald L. Newcomb Collection

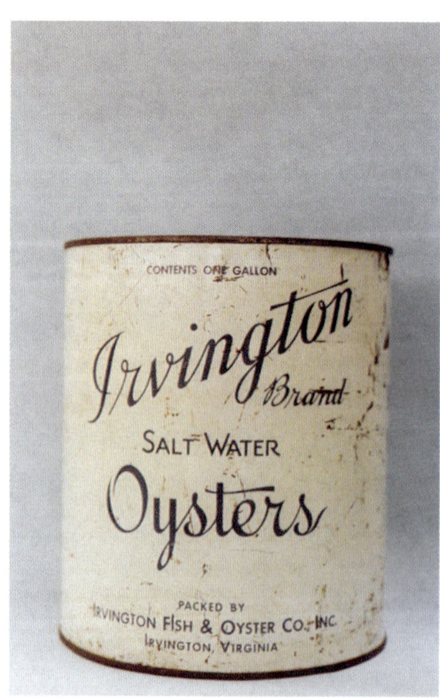

Irvington Fish & Oyster Co., Irvington, VA***
Irvington Brand, Gallon, VA 116
Bill and Steve Dorrell Collection

Cornwell Seafood Co., Irvington, VA***
Gallon, VA 88
Ronald L. Newcomb Collection

Crockett & Jones, Irvington, VA**
Gallon, VA 668
Carlton and Mary Riggin Collection

Warwick & Ashburn, Weems, VA***
W. & A. Brand, Gallon, VA 133
Ronald L. Newcomb Collection

Cornwell Sea Food Co., Irvington, VA***
Cornwell's Brand, Gallon, VA 272
Carlton and Mary Riggin Collection

Abbott Bros., Weems, VA***
Gallon, VA 103
Bill and Steve Dorrell Collection

Harding Seafood Co., Weems, VA***
Choice of Chesapeake Brand, Gallon, VA 120
Bill and Steve Dorrell Collection

Oyster World, Inc., Weems, VA**
Oyster World Brand, Gallon, VA 504
Carlton and Mary Riggin Collection

E.I. Webb & Co., Weems, VA***
Dutch Cove Brand, Gallon, VA 88
Mike and Eva Pinder Collection

W.F. Morgan & Sons, Weems, VA**
Morgan Brand, Gallon, VA 92
Carlton and Mary Riggin Collection

E.I. Webb & Co., Weems, VA***
Moonlight Bay Brand, Gallon, VA88
Carlton and Mary Riggin Collection

W.F. Morgan & Sons, Weems, VA***
Gallon
Carlton and Mary Riggin Collection

W.F. Morgan & Son, Weems, VA**
Gallon, VA 92
Carlton and Mary Riggin Collection

E.I. Webb & Co., Weems, VA***
Moonlight Bay Brand 12 Ounce, VA 88,
Drinking Cup
Bill and Steve Dorrell Collection

W.F. Morgan & Sons, Weems, VA***
Westbrook Brand, Gallon
Courtesy of Harris Crab House

J.S. Norris & Son, Weems, VA***
Gallon, VA 83
Ronald L. Newcomb Collection

Weems Seafood, Weems, VA***
Winstead Brand, Gallon, VA 308
Randy Shreck Collection

Rappahannock Oyster Co., Kilmarnock, VA***
Indian Creek Brand, Gallon, VA 210
Ronald L. Newcomb Collection

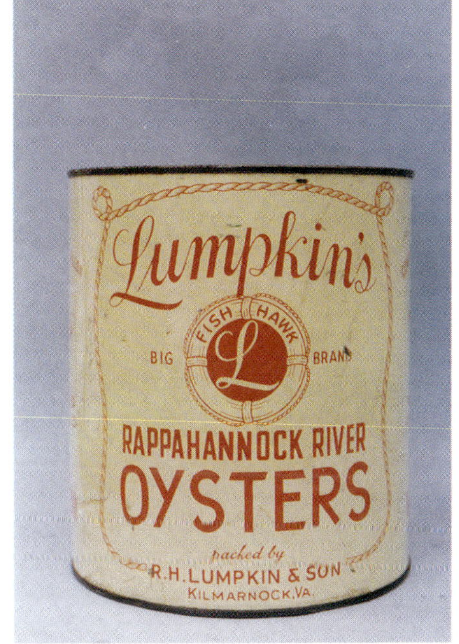

R.H. Lumpkin & Son, Kilmarnock, VA****
Fish Hawk Brand, Gallon
Carlton and Mary Riggin Collection

Oscar Ashburn & Son, Weems, VA****
Guiding Light Brand, Gallon, VA 594
Carlton and Mary Riggin Collection

Rappahannock Oyster Co., Kilmarnock, VA***
Bluff Point Brand, Gallon, VA 210
Carlton and Mary Riggin Collection

Ellery Kellum, Weems, VA***
Kellum Brand, Gallon, VA 15
Carlton and Mary Riggin Collection

Silver Sea Oyster, Inc., White Stone, VA**
Silver Sea Brand, Gallon, VA 90
Carlton and Mary Riggin Collection

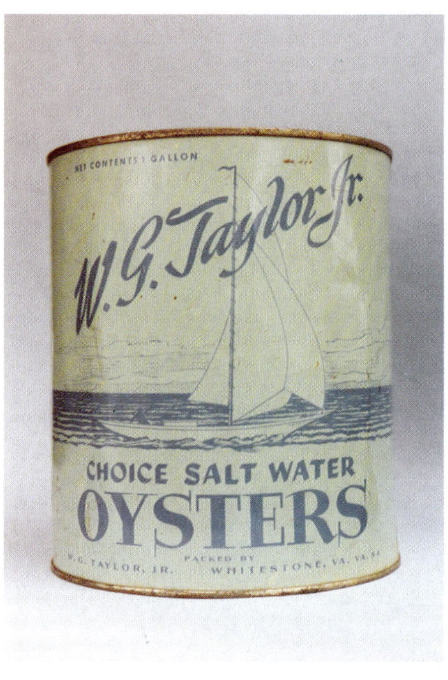

W.G. Taylor Jr., Whitestone, VA***
Gallon, VA 84
Ronald L. Newcomb Collection

Hughlett Seafood, Kilmarnock, VA**
Hughlett Point Brand, Gallon, VA 267
Carlton and Mary Riggin Collection

Nansemond Adams Oyster Co., Crittenden, VA***
Bleakhorn Brand, Gallon, VA 22
Carlton and Mary Riggin Collection

Irvington Packing Co., Whitestone, VA***
Old Dominion Brand, Gallon, VA 93
Carlton and Mary Riggin Collection

H. K. Billups, Motorun, VA****
Billups' Brand, Gallon, VA 464
Bill and Steve Dorrell Collection

B. M. Bunting's Oyster House, Gloucester, VA***
B & M Brand, Gallon, VA 387
Carlton and Mary Riggin Collection

Gloucester Seafood Packing Co., Bena, VA****
Duke of Gloucester Brand, Gallon, VA 394
Carlton and Mary Riggin Collection

H. K. Billups, Matthews, VA***
Billups' Brand, Gallon, VA 464
Carlton and Mary Riggin Collection

Geo. A. Philpotts, Mobjack, VA**
Gallon, VA 237
Carlton and Mary Riggin Collection

Tidewater Seafood Co., Bena, VA***
Pint, VA 87
Bill and Steve Dorrell Collection

Cooks Oyster Co., Bena, VA***
York Brand, Gallon
Courtesy of Harris Crab House

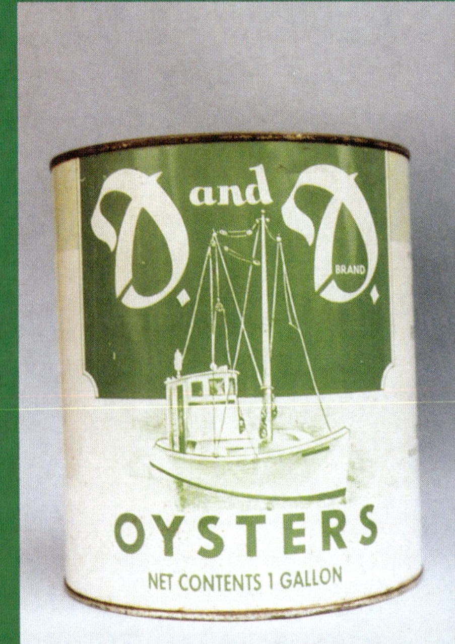

Dean Seafood, Montross, VA***
D. & D. Brand, Gallon, VA 124
Ronald L. Newcomb Collection

The National Packing Co., White Stone, VA***
Banner Brand, Pint, Va 93
Mike and Eva Pinder Collection

Cook's Seafood Co., Bena, VA***
Cook's Brand, Gallon, VA 607
Carlton and Mary Riggin Collection

Allen's Oyster House, Coles Point, VA**
Allen's Brand, Pint, VA 13
Mike and Eva Pinder

DELAWARE

Cropper Oyster Co., Dagsboro, DE****
Gallon, DE 26
Ronald L. Newcomb Collection

Tignor Oyster Packing Co., Bowers, DE*****
North Crystal Brand, Gallon, DE 23
Ronald L. Newcomb Collection

J. B. Robinson & Co., Seaford, DE****
Gallon, DE 2
Ronald L. Newcomb Collection

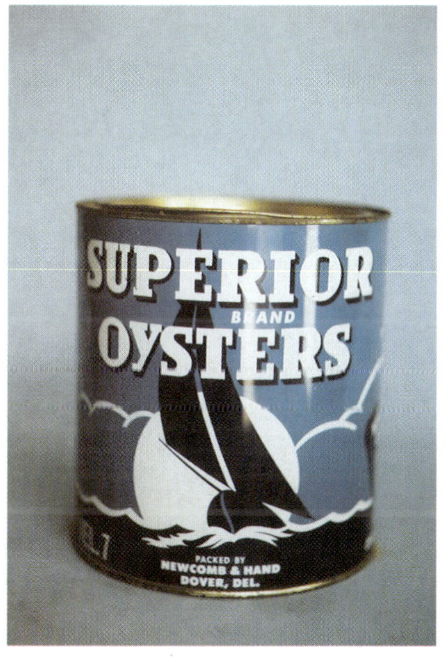

Newcomb & Hand, Dover, DE**
Superior Brand, Gallon, DE 7

Booth Fisheries Co., Dover, DE**
Gallon, DE 26
Carlton and Mary Riggin Collection

J. B. Robinson & Co., Seaford, DE****
Pint, DE 2
Charles & Jane Wilkins Collection

Allen Kirkpatrick & Co., Rehoboth, DE****
Kirkpatrick's Brand, Gallon, DE 3
Ronald L. Newcomb Collection

Allen Kirkpatrick & Co., Rehoboth, DE***
Kirkpatrick's Brand, Gallon, DE 36
Courtesy of Harris Crab House

Allen Kirkpatrick & Co., Rehoboth, DE**
Kirkpatrick's Brand, Gallon, DE 3
Carlton and Mary Riggin Collection

Allen Kirkpatrick & Co., Dover, DE****
Kirkpatrick's Brand, Pint, DE 3
Bill and Steve Dorrell Collection

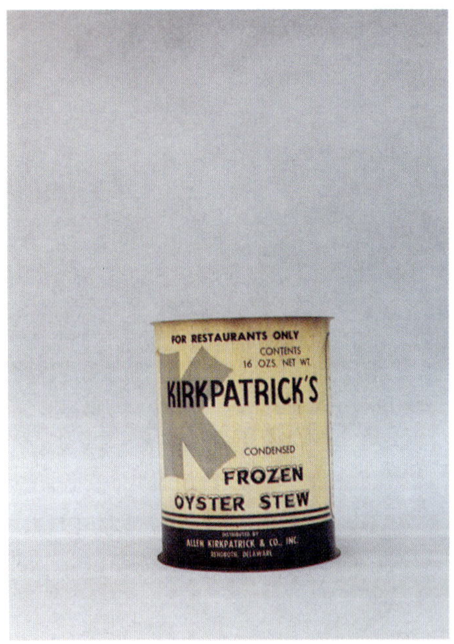

Allen Kirkpatrick & Co., Rehoboth, DE**
Kirkpatrick's Brand, 16 Ounces
Courtesy of Harris Crab House

Allen Kirkpatrick & Co., Rehoboth, DE**
Kirkpatrick's Brand, Gallon, DE 3
Carlton and Mary Riggin Collection

Allen Kirkpatrick & Co., Rehoboth, DE**
Kirkpatrick's Brand, Half Gallon, FL 346 RP
Butch and Jackie Cheezum Collection

PENNSYLVANIA

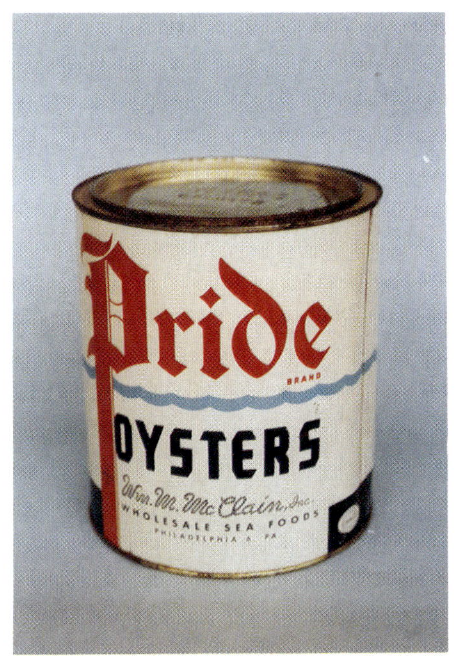

Wm. M. McClain, Philadelphia, PA****
Pride Brand, Gallon, PA 8
Courtesy of Black Swan Antiques

Edgar T. Hill, Philadelphia, PA****
Gallon
Carlton and Mary Riggin Collection

Delaware Seafood Co., Philadelphia, PA***
Oyster Boat Brand, Gallon, PA 336
Carlton and Mary Riggin Collection

Jarrell & Rea, Pittsburgh, PA***
Silver Sea Brand, Pint, MD 81
Carlton and Mary Riggin Collection

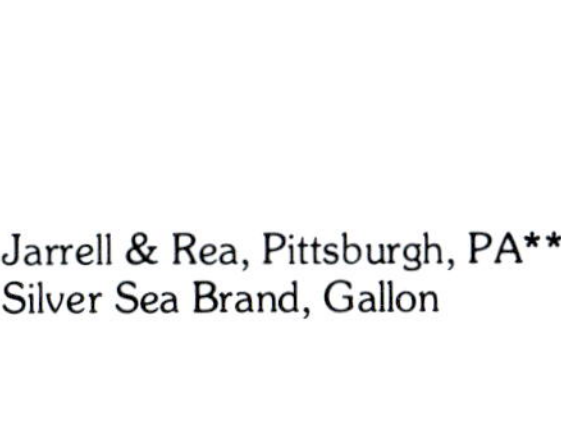

Jarrell & Rea, Pittsburgh, PA**
Silver Sea Brand, Gallon

R. W. Strickler, York, PA***
Strickler's Brand, 12 Ounces

Bierman's, York, PA****
Quart
Carlton and Mary Riggin Collection

R. W. Strickler, York, PA**
Srickler's Brand, 2 Pint Cans, MD 202

Morrison & McCluan, Pittsburgh, PA***
Shamrock Brand, Pint, MD 51
Ronald L. Newcomb Collection

NEW JERSEY

Port Norris Oyster Co., Port Norris, NJ***
Coast-Pact Brand, Gallon, NJ 1
Carlton and Mary Riggin Collection

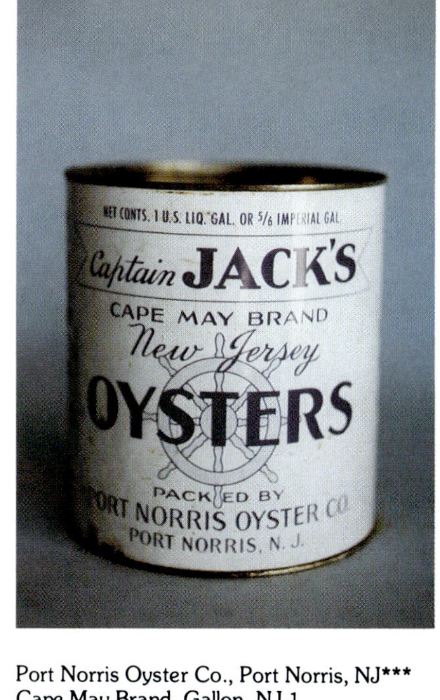

Port Norris Oyster Co., Port Norris, NJ***
Cape May Brand, Gallon, NJ 1
Carlton and Mary Riggin Collection

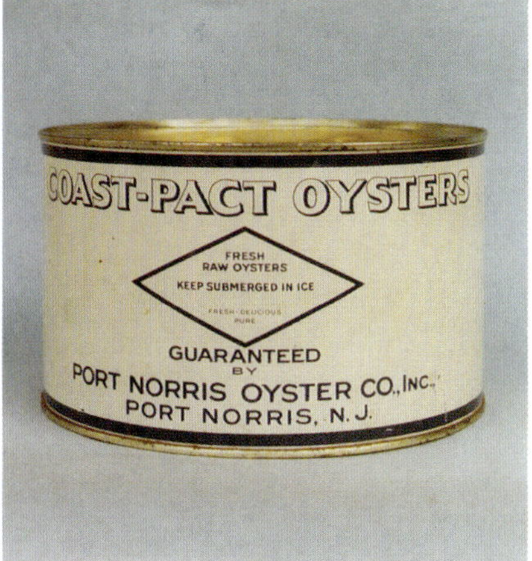

Port Norris Oyster Co., Port Norris, NJ***
Coast-Pact Brand, Half Gallon, NJ 1
Butch and Jackie Cheezum Collection

Port Norris Oyster Co., Port Norris, NJ**
Coast-Pact Brand, Gallon, NJ 1
Carlton and Mary Riggin Collection

Reed & Reed, Port Norris NJ**
Gallon, NJ 210
Carlton and Mary Riggin Collection

Phillips Seafood Packing Co., Port Norris, NJ***
Jersey Cape Brand, Gallon, NJ 987
Carlton and Mary Riggin Collection

Planters Packing Company, Port Norris, NJ***
Gallon, NJ 2
Ronald L. Newcomb Collection

Peterson Packing Co., Port Norris, NJ**
Gallon, NJ 16
Carlton and Mary Riggin Collection

Geo. A. McConnell, Port Norris, NJ**
New Jersey Brand, Gallon, NJ 557
Carlton and Mary Riggin Collection

Peterson Packing Co., Port Norris, NJ***
Miah-Maull Brand, Half Gallon, NJ 16
Ronald L. Newcomb Collection

Garden State Oyster Co., Port Norris, NJ****
Garden State Brand, Gallon, PA 29
Ronald L. Newcomb Collection

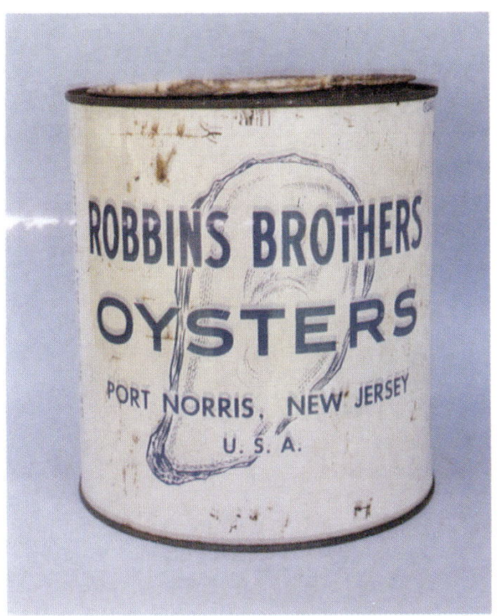

Robbins Brothers, Port Norris, NJ***
Gallon, NJ 6
Carlton and Mary Riggin Collection

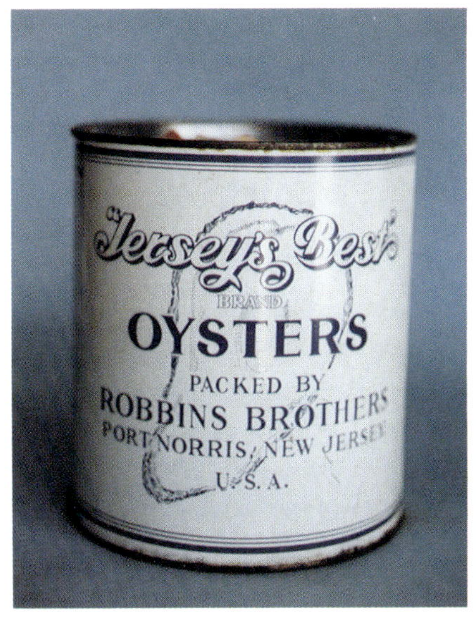

Robbins Brothers, Port Norris, NJ***
Jersey's Best Brand, Gallon, NJ 6
Carlton and Mary Riggin Collection

DuBois & Son's Co., Bivalve, NJ***
Jersey Brand, Pint, NJ 176
Ralph and Betty Tull Collection

Robbins Brothers, Port Norris, NJ**
Jersey's Best Brand, Gallon, NJ 6
Carlton and Mary Riggin Collection

Newcomb & Hollinger Oyster Co., Port Norris, NJ****
Quality Brand, Gallon, NJ 4

Jeffries Oyster Farms, Bivalve, NJ***
Jeffries Brand, Gallon, NJ 4
Ralph and Betty Tull Collection

Robbins Brothers, Port Norris, NJ***
Half Gallon, NJ 6
Butch and Jackie Cheezum Collection

Unmarked, Packed by Stowman Bros.****
Maurice River, NJ
High Tide Brand, NJ 5
Ronald L. Newcomb Collection

Stowman Brothers, Maurice River, NJ***
Cape May Brand, Half Gallon, NJ 5
Bill and Steve Dorrell Collection

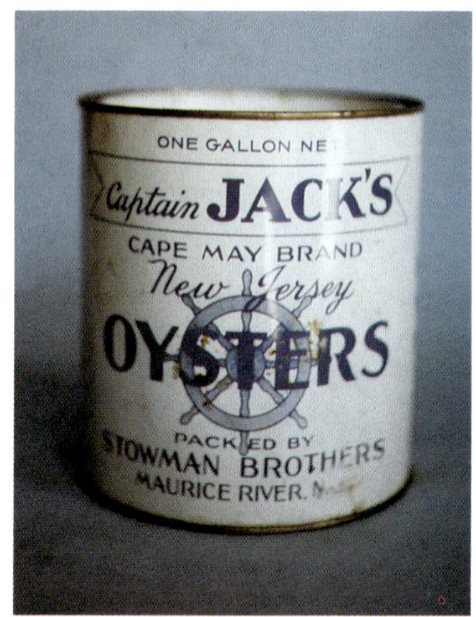

Stowman Brothers, Maurice River, NJ***
Cape May Brand, Gallon, NJ 5
Carlton and Mary Riggin Collection

Back of Above Can

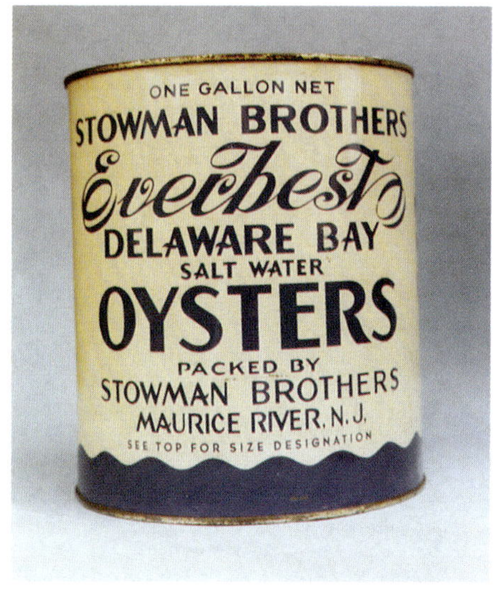

Stowman Brothers, Maurice River, NJ***
Ever Best Brand, Gallon, NJ 5
Ronald L. Newcomb Collection

Bivalve Seafood Co., Bivalve, NJ***
Famous Bivalve Brand, Half Gallon, NJ 902
Courtesy of Harris Crab House

Bivalve Packing Co., Bivalve, NJ**
Treat From the Deep Brand, Gallon, NJ 19
Carlton and Mary Riggin Collection

Bivalve Seafood Co., Bivalve, NJ***
Famous Bivalve Brand, Gallon, NJ 902
Donald C. Bell Collection

Fogg & Stowman, Bivalve, NJ***
Pint, NJ 40
Ronald L. Newcomb Collection

Joseph N. Fowler & Son, Leesburg, NJ**
Gallon, NJ 7
Courtesy of Harris Crab House

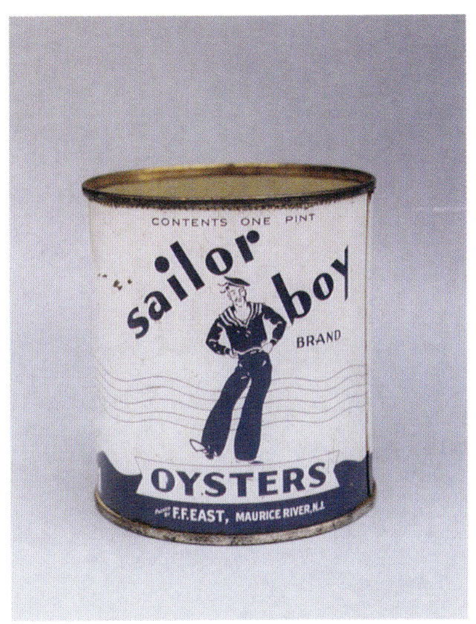

F.F. East, Inc., Maurice River, NJ***
Sailor Boy Brand, Pint, NJ 11
Ronald L. Newcomb Collection

Dill's Seafood, Bridgeton, NJ***
Dill's Brand, Gallon, NJ 17
Carlton and Mary Riggin Collection

E.C. DuBois & Co., Bivalve, NJ****
Jersey Brand, Tall Quart, NJ 176
Ronald L. Newcomb Collection

F.F. East, Inc., Maurice River, NJ****
Sailor Boy Brand, Tall Quart, NJ 11
Ronald L. Newcomb Collection

Greenwich Oyster Co., Greenwich, NJ***
Eastpoint Brand, Gallon, NJ 17

East Point Oyster Co., Cape May Court House, NJ*****
Cape May Salts Brand, Gallon, NJ 33
Carlton and Mary Riggin Collection

Greenwich Oyster Co., Greenwich, NJ**
Gallon, NJ 17
Carlton and Mary Riggin Collection

Lester & Toner, Greenport, NY**
Seapure Brand, Gallon, NY 4
Carlton and Mary Riggin Collection

Geo. M. Still, New York, NY***
Diamond Point, Gallon, NY 24
Butch and Jackie Cheezum Collection

Lester & Toner, Greenport, NY**
Seapure Brand, Half Gallon
Butch and Jackie Cheezum Collection

Bluepoints Co., West Sayville, NY***
Sealshipt Brand, Half Gallon, NY 24
Carlton and Mary Riggin Collection

J. & J.W. Elsworth Co., New York, NY***
Quality First Brand, Gallon, NY 3
Courtesy of Harris Crab House

Geo. Thompson & Son, Greenport, NY***
Sea X Cross Brand, Gallon, NJ 19
Donald C. Bell Collection

Lid From Can Above

J. & J.W. Elsworth Co., Montauk Point, NY****
Red Cross Brand, Gallon, NY 3
Courtesy of Harris Crab House

Shelter Island Oyster Co., Greenport, NY**
Shelter Island Brand, 6 Sizes, Gallon to 12 Ounces, NY 133
Donald C. Bell Collection

Montauk Seafood Co., New York, NY**
Montauk Brand, Gallon, VA 45

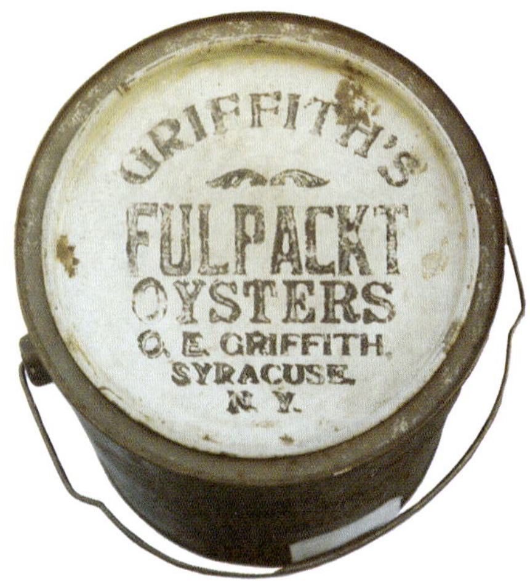

O.E. Griffiths, Syracuse, NY
Picture of Lid
Courtesy of Burt Brooks

Schacht Seafoods, New York, NY***
Viking Brand, Gallon, DE 3
Ronald L. Newcomb Collection

Oyster Bay Oyster Co., Oyster Bay, NY***
Seawanhaka Brand, Gallon, NYS 1
Donald C. Bell Collection

Ernest C. Holmes, Waterloo, NY****
Full Packt Brand, Gallon, MD 39
Carlton and Mary Riggin Collection

CONNECTICUT

Frederick F. Lovejoy, East Norwalk, CT***
White Rock and Grassy Hammock Brand, Gallon, CT 6
Randy Shreck Collection

H. C. Rowe & Co., New Haven CT***
Rowe's Brand, Gallon, CT 2

Tallmadge Bros., South Norwalk, CT**
Cedar Point Brand, Gallon, CT 10
Donald C. Bell Collection

Tallmadge Bros., South Norwalk, CT***
Cedar Point Brand, Gallon, CT 10

RHODE ISLAND

Warren Oyster Co., Warren RI***
New England Brand, Gallon, RI 5
Carlton and Mary Riggin Collection

Warren Oyster Co., Warren, RI***
New England Brand, Gallon, RI 5
Carlton and Mary Riggin Collection

Narragansett Bay Oyster Co., Warren, RI***
Sea Acre Brand, Gallon, RI 1

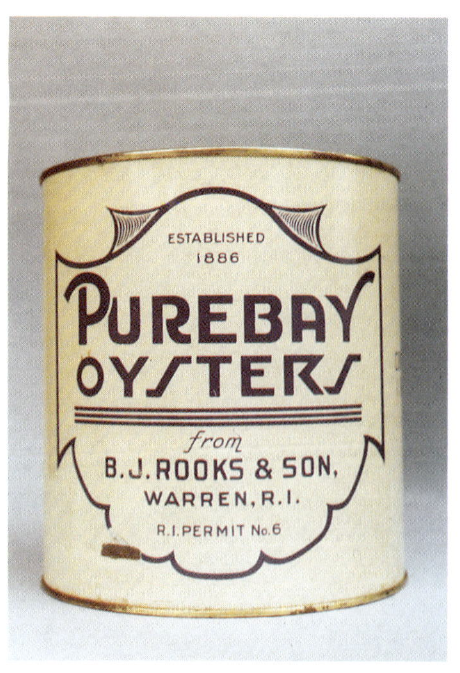

B. J. Rooks & Son, Warren, RI***
Purebay Brand, Gallon, RI 6
Ronald L. Newcomb Collection

MASSACHUSETTS

United Shellfish Company, Ipswich, MA***
Famous Brand, Gallon, MA 175
Bill and Steve Dorrell Collection

Whitman, Ward & Lee Co., Boston, MA****
Harvard Brand, Gallon

40 Fathom Fish Co., Boston, MA***
Pint, VA 214

Ipswich Shellfish Company, Ipswich MA***
Blue Seal Brand, Gallon, MD 169
Carlton and Mary Riggin Collection

B & S Fisheries, Fall River, MA***
Queen's Choice Brand, Gallon, MD 606

Ralph H. Ranagan, Boston MA***
Salt Air Brand, Gallon, NJ 210
Donald C. Bell Collection

CANADA

Quebec United Fishermen, Montreal**
Bluecold Brand, Gallon, VA 45

Quebec United Fishermen, Montreal***
La Mariniere Brand, Gallon, VA 437
Ronald L. Newcomb Collection

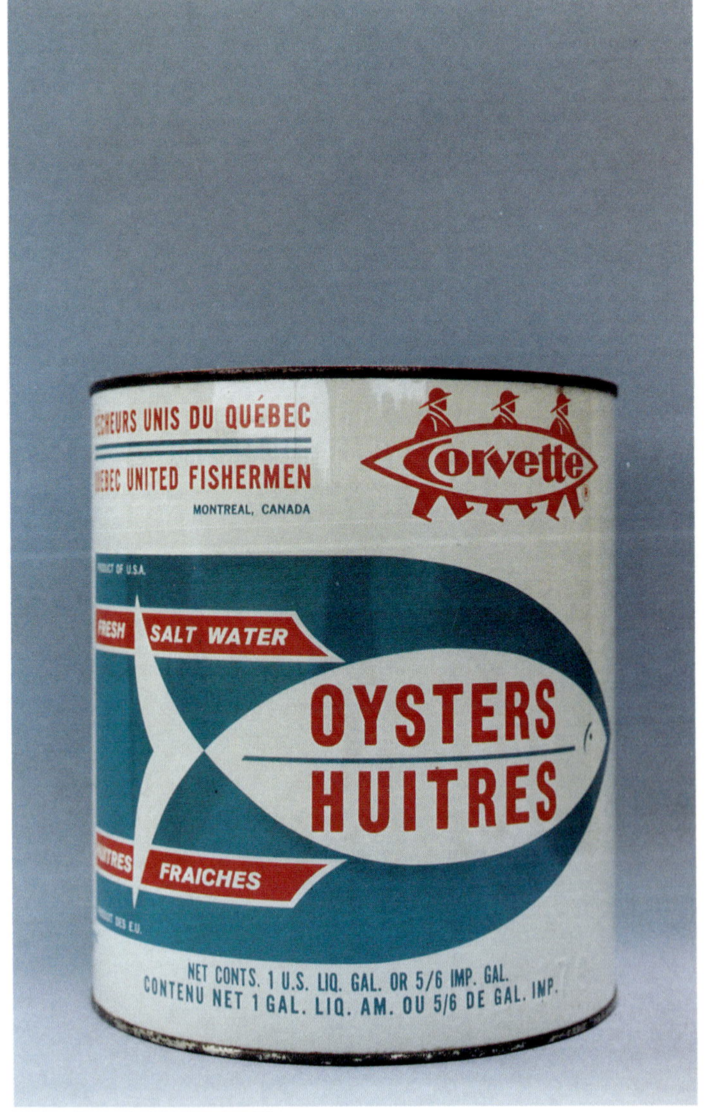

Quebec United Fishermen, Montreal**
Corvette Brand, Gallon, MD 176
Bill and Steve Dorrell Collection

NORTH CAROLINA

Englehard Shrimp, Fish and Oyster Co.,***
Englehard, NC
Chief Englehard Brand, Gallon, NC 142
Carlton and Mary Riggin Collection

Kinlaw Seafood Co., Shallotte Point, NC***
Gallon
Carlton and Mary Riggin Collection

J.S. Clifton, Shallotte, NC***
Clifton's Brand, Pint, NC 9
Bill and Steve Dorrell Collection

Packer Unknown**
Pint, NC 206
Mike and Eva Pinder Collection

Clayton Fulcher Seafood Co., Atlantic, NC***
Diamond Shoals Brand, Gallon, NC 8

C.B. Jeannette, Swan Quarter, NC****
Rose Bay Brand, Gallon, NC 165
Courtesy of Harris Crab House

Green's Oyster Co., Shallotte, NC***
Green's Brand, Gallon, NC 32
Ralph and Betty Tull Collection

GEORGIA

L. P. Maggioni & Co., Savannah, GA**
Daufuski Brand, Gallon, SC 38
Carlton and Mary Riggin Collection

Capital Fish Co., Atlanta, GA***
Capital Brand, Pint, MD 45
Ronald L. Newcomb Collection

J. S. Graves, Jr., Bluffton, SC***
Bluffton Brand, Pint
Ronald L. Newcomb Collection

SOUTH CAROLINA

L. P. Maggioni & Co., Savannah, GA**
Daufuski Brand, Pint, SC 9
Carlton and Mary Riggin Collection

Glines Seafood, Awendaw, SC***
Bull Bay Brand, Gallon, SC 29
Courtesy of Harris Crab House

A. S. Varn & Son, Savannah, GA***
Pin Point Brand, Half Pint, GA 9
Courtesy of Harris Crab House

H.D. Hutchins, Hilton Head, SC***
Hilton Head Brand, Pint, SC 138
Carlton and Mary Riggin Collection

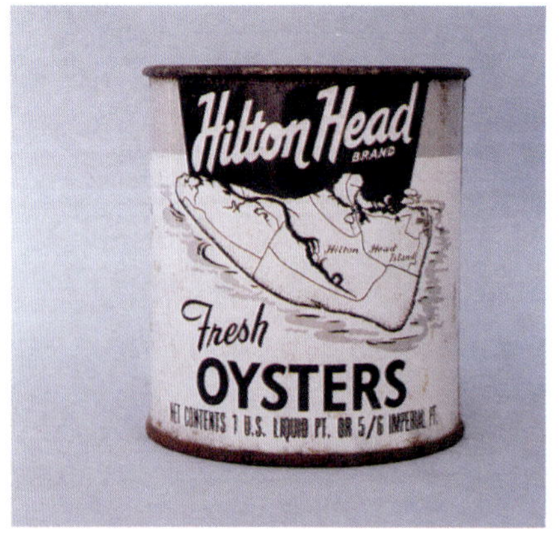

FLORIDA

Kirvin Bros. Sea Food, Apalachicola, FL****
Apalachicola Bay Brand, Gallon, FL 53
Carlton and Mary Riggin Collection

Florida Bahama Seafood Co., Boynton Beach, FL*
Florida Bahama Brand, 8 Ounces, 12 Ounces, 16 Ounces

Gulf's Best Food, Bayou La Batre, AL***
Gulf's Best Brand, Gallon, AL 4
Carlton and Mary Riggin Collection

ALABAMA

Seaman Fish Co., Bayou La Batre, AL***
Seaman Brand, Gallon, AL 97
Carlton and Mary Riggin Collection

2 New Glass Containers, Alabama, AL 2 & 91 *
Carlton and Mary Riggin Collection

Bon Secour Fisheries, Bon Secour, AL**
Nelson's Brand, Gallon, AL 49
Carlton and Mary Riggin Collection

Jeff Johnson, Geron Bay, AL***
Jeff Johnson's Brand, Gallon, AL 29
Ronald L. Newcomb Collection

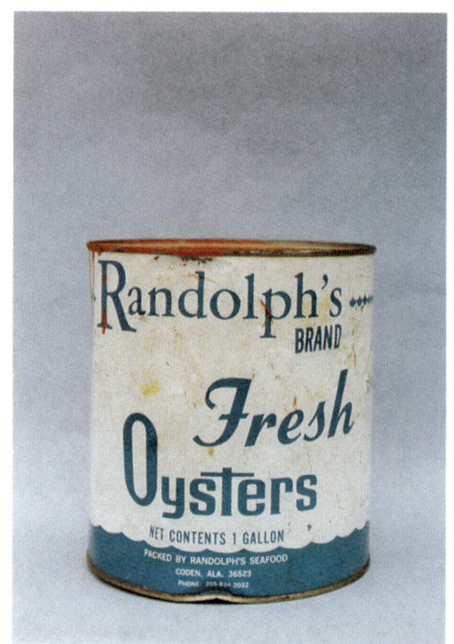

Randolph's Seafood, Coden, AL***
Randolph's Brand, Gallon, AL 54
Courtesy of Harris Crab House

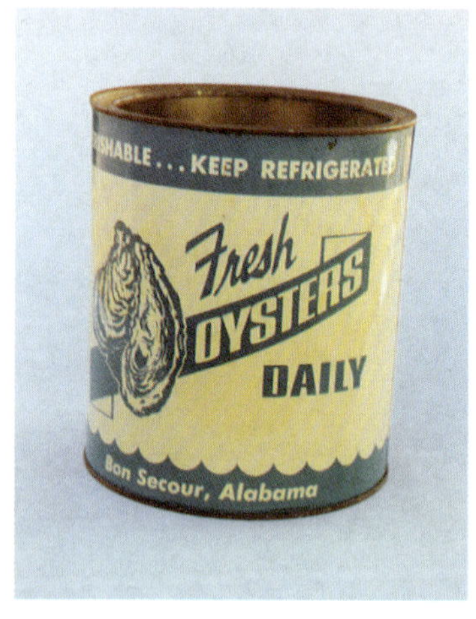

Bon Secour Fisheries, Bon Secour, AL***
Carlton and Mary Riggin Collection

Randolph's Seafood, Coden, AL***
Capt. Collier's Brand, Gallon, AL 54
Courtesy of Harris Crab House

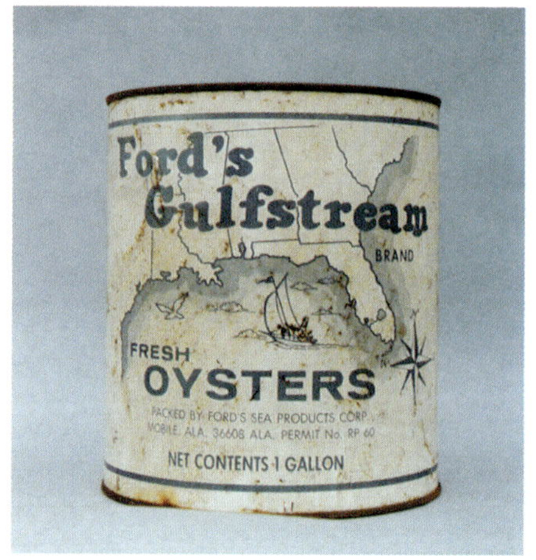

Ford's Sea Products, Mobile, AL***
Ford's Gulfstrean Brand, Gallon, AL 60
Courtesy of Harris Crab House

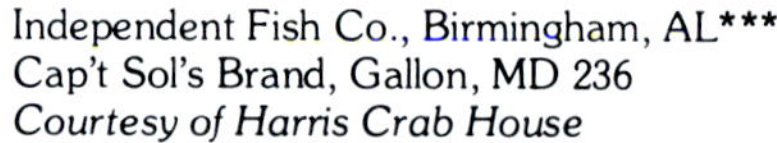
Independent Fish Co., Birmingham, AL***
Cap't Sol's Brand, Gallon, MD 236
Courtesy of Harris Crab House

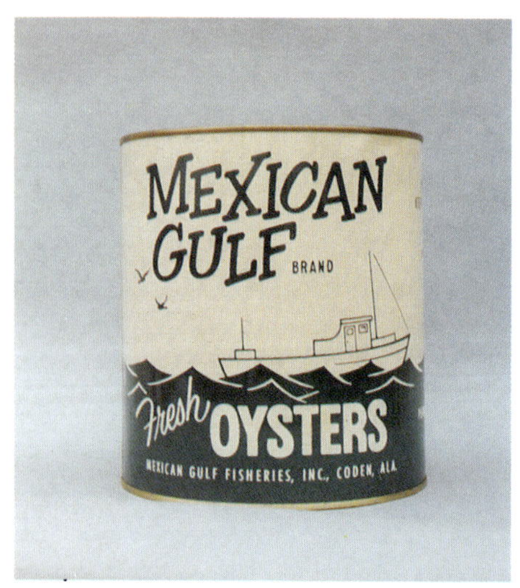

Mexican Gulf Fisheries, Coden, AL***
Mexican Gulf Brand, Gallon, AL 14
Courtesy of Harris Crab House

MISSISSIPPI

Shemper Seafood Co., Biloxi, MS**
Shemper's Brand, 12 Ounces, MS 151
Courtesy of Harris Crab House

Bennett's Seafood Co., Pass Christian, MS***
Bennetts's Brand, Gallon, MS 146
Courtesy of Harris Crab House

DeJean Packing Co., Biloxi, MS****
DeJean's Brand, Gallon, MS 6
Bill and Steve Dorrell Collection

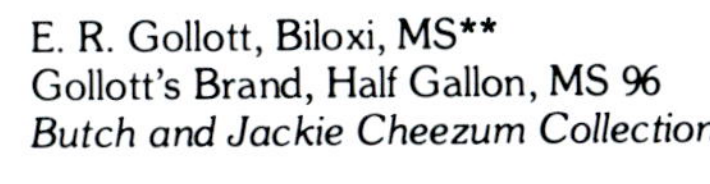

E. R. Gollott, Biloxi, MS**
Gollott's Brand, Half Gallon, MS 96
Butch and Jackie Cheezum Collection

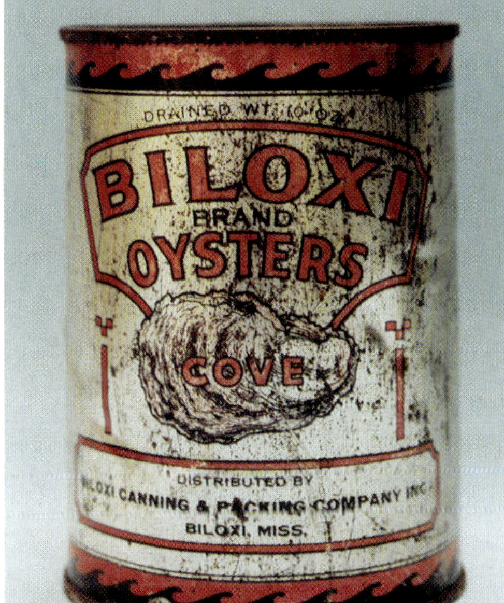

Biloxi Canning & Packing Co., Biloxi, MS****
Biloxi Brand, 10 Ounces
Bill and Steve Dorrell Collection

C.F. Gollott & Son Seafood Co., Biloxi, MS***
Gollott's Brand, Pint, MS 79
Ronald L. Newcomb Collection

E.R. Gollott, Biloxi, MS**
Capt'n Gollott Brand, Gallon, MS 206
Courtesy of Harris Crab House

LOUISIANA

Monjure Seafood, New Orleans, LA**
Gallon, LA 178
Courtesy of Harris Crab House

J. Kambur Co., New Orleans, LA***
Kambur's Brand, Gallon, LA 83
Courtesy of Harris Crab House

Martina & Martina, New Orleans, LA**
M M Brand, Pint, LA 18
Courtesy of Harris Crab House

Capt. Michael's Oysters, New Orleans, LA**
Gallon, LA 78
Carlton and Mary Riggin Collection

Golden Meadow Oyster House, Golden Meadow, LA**
Captain Pitre's Best Brand, Gallon, LA 53
Courtesy of Harris Crab House

Roland White Oyster Co., St Bernard, LA**
Roland White's Brand, Gallon, LA 532
Carlton and Mary Riggin Collection

TEXAS

Pelican Oyster & Fish Co., New Orleans, LA*****
Pelican Brand, Gallon, LA 8
Bill and Steve Dorrell Collection

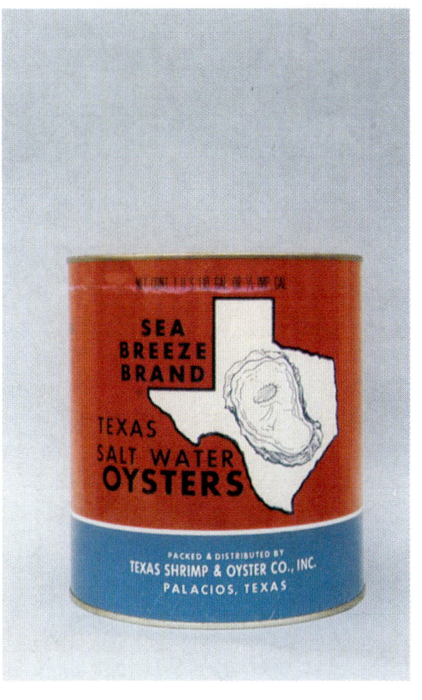

Texas Shrimp & Oyster Co., Palacios, TX***
Sea Breeze Brand, Gallon, TX 122SP
Courtesy of Harris Crab House

Joe Grasso & Son, Galveston, TX***
Grasso Brand, GallonTX 220
Courtesy of Harris Crab House

DeSanka Oyster House, Amite, LA**
Louisiana Bayou Brand, 9 ½ Ounces, LA 70
Courtesy of Harris Crab House

OHIO

David Davies, Columbus OH***
Double D D Brand, Pint, MD 220
Carlton and Mary Riggin Collection

Big Bear Stores Co., Columbus, OH**
Fresh Compass Brand, Pint
Ronald L. Newcomb Collection

Lynch Fish Co., Cincinnati, OH***
East Coast Brand, Gallon

The New Fisheries Co., Cincinnati, OH****
Very Best Brand, Tall Quart
Ronald L. Newcomb Brand

Schneier's Seafoods, Akron OH**
Schneier's Brand, Pint
Mike and Eva Pinder Collection

Tracy & Avery Co., Mansfield, OH***
Westbrook Brand, Gallon, MD 224
Carlton and Mary Riggin Collection

Central Fruit Co., Mansfield, OH***
Some Oysters Brand, Gallon
Joe Sechrist Collection

Kroger Co., Cincinnati, OH**
Fres-shore Brand, Pint

Geo. H. Thomas, Cincinnati, OH****
Thomas Brand, Gallon MD 81
Ronald L. Newcomb Collection

TENNESSEE, NEBRASKA, IOWA

Anderson Fish & Oyster Co., Nashville, TN****
Quality A Brand, Pint
Carlton and Mary Riggin Collection

The King Cole Company, Omaha, NE***
King Cole Brand, Gallon
Ronald L. Newcomb Collection

Cedar-Rapids Commission Co., Cedar Rapids, IA***
Quality Brand, Gallon, MD 208

Wahkonsa Packing Co., Fort Dodge, IA****
Wahkonsa Brand, Gallon, MD 45
Mike and Eva Pinder Collection

MICHIGAN

R. F. Brown Sea Food Co., Lansing, MI**
North Crystal Brand, Gallon, VA 283

R. F. Brown Sea Food Co., Lansing, MI**
Sailor Boy Brand, Gallon, MD 21

Sterling Oyster Co., Detroit, MI**
Keystone Brand, 12 Ounces, MD 81
Courtesy of Harris Crab House

R. F. Brown Sea Food Co., Lansing, MI**
Cream O' Sea Brand, Gallon, VA 277

Sterling Oyster Co., Detroit, MI***
Half Gallon
Carlton and Mary Riggin Collection

ILLINOIS

Chas. W. Triggs Co.,****
Continental Can Co., Chicago, IL
Sea Gull Brand, One Fifth Gallon
Ronald L. Newcomb Collection

Booth Fisheries Co., Chicago, IL****
Booth Brand, Gallon, VA 197
Carlton and Mary Riggin Collection

Booth Fisheries Co., Chicago, IL***
Booth Brand, Pint, MD 166
Mike and Eva Pinder Collection

Booth Fisheries Co., Chicago, IL***
Booth Brand, Gallon, MD 197
Carlton and Mary Riggin Collection

Edwin M. Plitt & Son, Chicago, IL***
Sailor Girl Brand, Pint, VA 437
Bill and Steve Dorrell Collection

WISCONSIN, COLORADO, WASHINGTON

A.F. Schwahn & Sons, Eau Claire, WI***
Schwahn's Brand, Gallon, NJ 5
Ronald L. Newcomb Collection

United Oyster Co., Seattle WA***
Pacific Oysters Brand, Gallon
Carlton and Mary Riggin Collection

Olympia Oyster Co., Shelton, WA**
Scots Point Brand, 12 Ounces,WA 7

M. G. Flint Mercantile Co., Denver CO****
F Brand, Gallon
Courtesy of Frank Speal, Jr.

United Oyster Producers Assoc., Seattle, WA**
Willapoint Brand, 12 Ounces

Haines Oyster Co., Seattle, WA**
Haines Brand, 12 Ounces

Olympia Oyster Co., Shelton, WA**
The Smiling Oyster Brand, 12 Ounces,
WA 7

CALIFORNIA

King's Choice Foods, Los Angeles, CA***
King's Choice Brand, Gallon
Ralph and Betty Tull Collection

STOCK CANS

Gallon Stock Can****

Half Gallon Lazy S Stock Can****

Gallon Stock Can***

Gallon Stock Can****
Mike and Eva Pinder Collection

Tall Quart Lazy S Stock Can****
Ronald L. Newcomb Collection

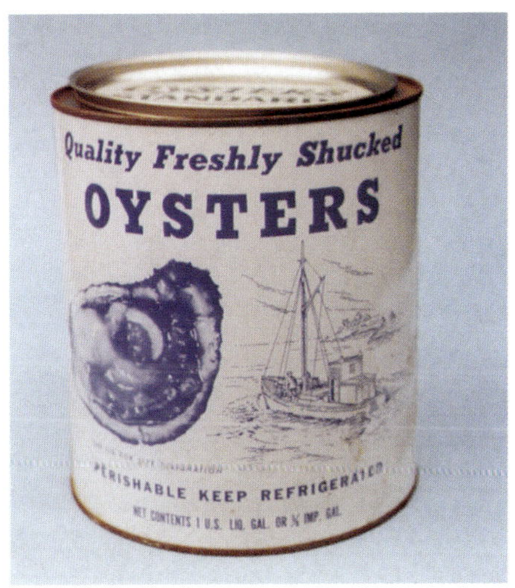

Gallon White Oyster Stock Can*

Gallon Lazy S Stock Can***
Carlton and Mary Riggin Collection

Group of Sailboat Stock Cans
Carlton and Mary Riggin Collection

SOLDERED CANS

Gallon Sailboat Stock Can****
Mike and Eva Pinder Collection

Gallon Flavor-Fresh Stock Can*

Eagle Packing Co., Baltimore, MD*****
Eagle Brand, Soldered Rectangular Can
Ronald L. Newcomb Collection

Louis Grebb Co., Baltimore, MD*****
Belle Brand, Soldered Rectangular Can

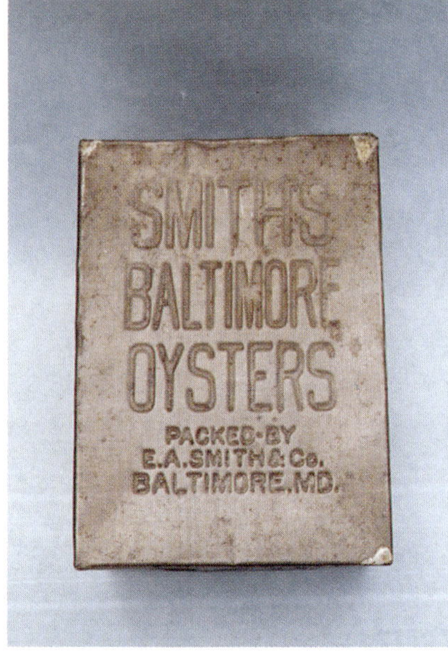

E.A. Smith & Co., Baltimore, MD*****
Smith's Brand, Soldered Rectangular Can
Gary and Sharon Campbell Collection

Thurston, Russell & Co., Annapolis. MD*****
Lightning Express Brand, Soldered Rectangular Can
Carlton and Mary Riggin Collection

PAPER LABEL CANS

Store & Bunnell, Baltimore, MD*****
Saddle Rock Brand, Quart
Soldered Rectangular Can
Ronald L. Newcomb Collection

C.H. Pearson & Co., Baltimore, MD*****
Peerless Brand, Soldered Rectangular Can

Moore & Brady, Baltimore, MD*****
Deep Sea Brand
Carlton and Mary Riggin Collection

Moore & Brady, Baltimore, MD*****
Deep Sea Brand
Ronald L. Newcomb Collection

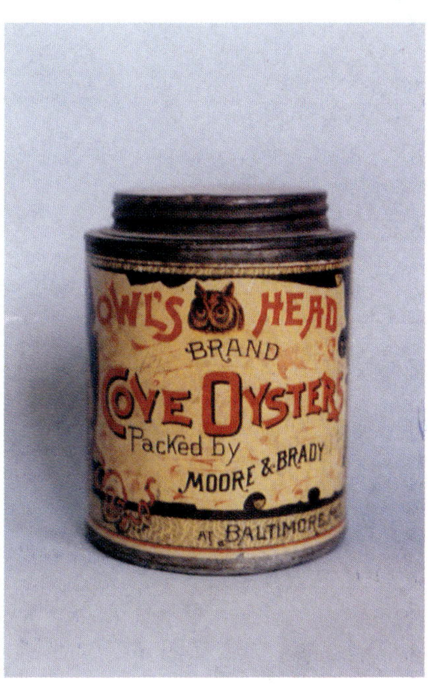

Moore & Brady, Baltimore, MD*****
Owl's Head Brand, Screw Top
Carlton and Mary Riggin Collection

Back of Above Can

H.S. Lanfair & Co., Baltimore, MD*****
Challenge Brand, Showing can opener in position
Ronald L. Newcomb Collection

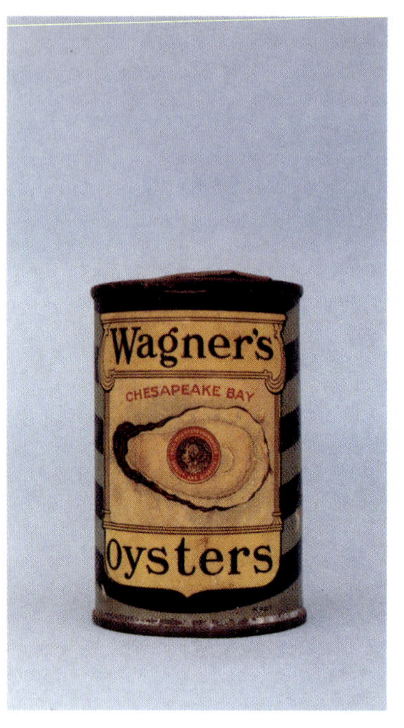

Martin Wagner Co., Baltimore, MD*****
Wagner's Brand, 5 Ounces

Baltimore & Chesapeake Oyster Co., Baltimore, MD*****
Deer Island Brand
Bill and Steve Dorrell Collection

Gibbs Preserving Co., Baltimore, MD*****
Bulls Head Brand, Quart
Gary and Sharon Campbell Collection

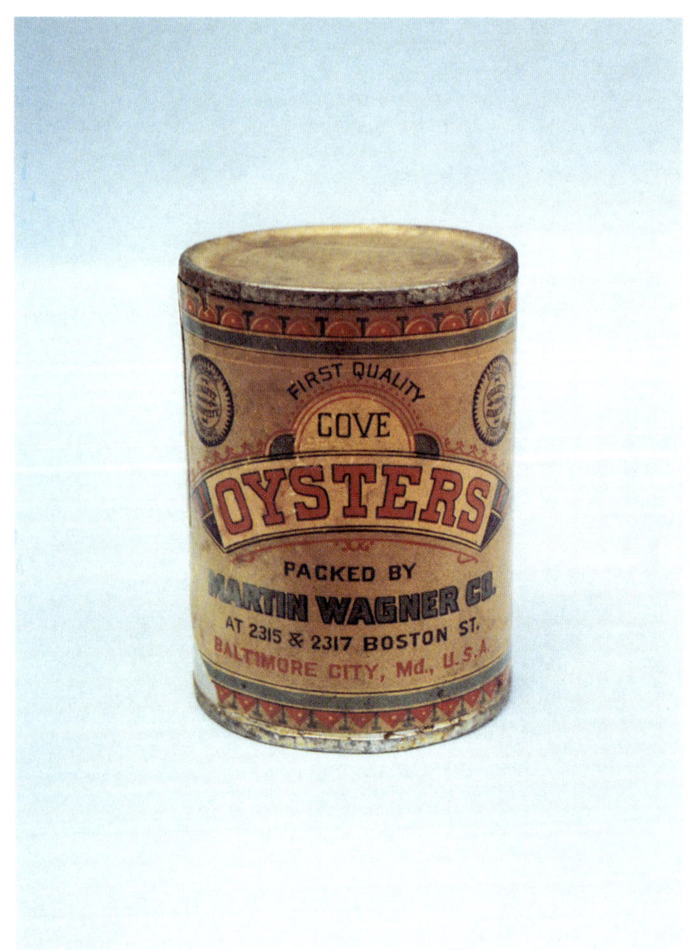

Martin Wagner Co., Baltimore, MD*****
Courtesy of Black Swan Antiques

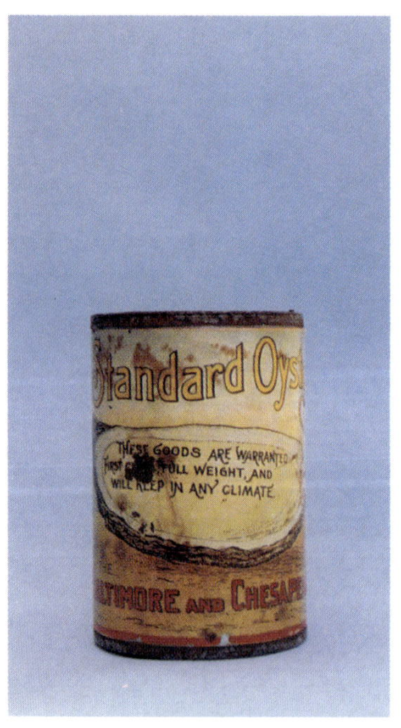

Back of Above Can

Back of Above Can

W.W. Boyer & Co., Baltimore, MD*****
Boyer's Brand
Ronald L. Newcomb Collection

J. Langrall & Bros., Baltimore, MD*****
Maryland Chief Brand, 5 Ounces
Bill and Steve Dorrell Collection

Wm. Taylor, Baltimore, MD*****
Packed for John Hunt & Co.
Joe Sechrist Collection

H. Beckwith & Co., Baltimore, MD*****
The Boss Brand
Ronald L. Newcomb Collection

Back of Above Can

Back of Above Can

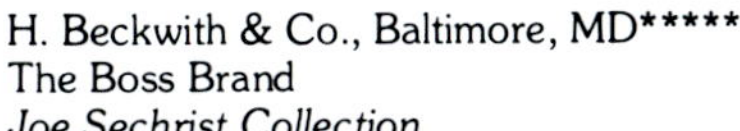
H. Beckwith & Co., Baltimore, MD*****
The Boss Brand
Joe Sechrist Collection

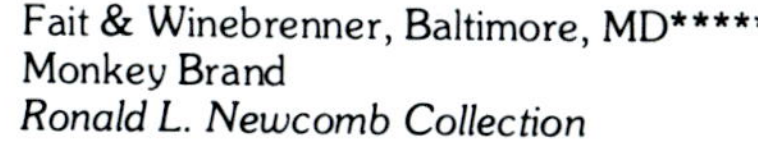
Fait & Winebrenner, Baltimore, MD*****
Monkey Brand
Ronald L. Newcomb Collection

F. H. Siewerd & Co., Baltimore, MD*****
Siewerd's Brand, 10 Ounces
Joe Sechrist Collection

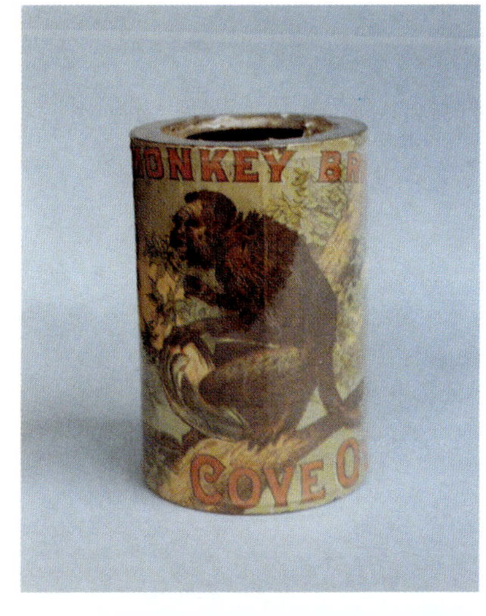

Packer Unknown, Baltimore, MD*****
Camel Brand
Ronald L. Newcomb Collection

C. A. Pearson Co., Baltimore, MD*****
Peerless Brand
Bill and Steve Dorrell Collection

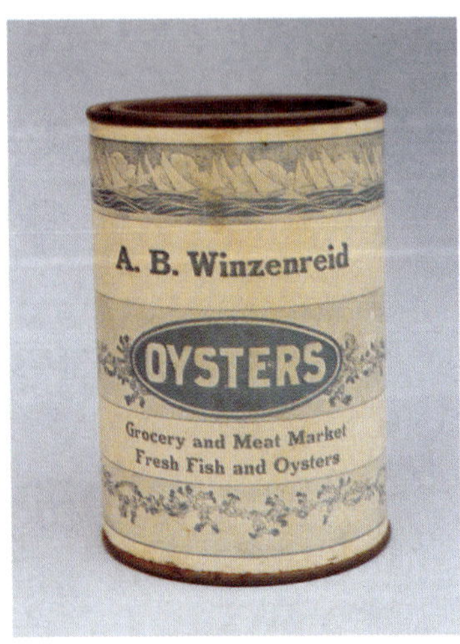

A. B. Winzenreid*****
Ronald L. Newcomb Collection

Packer Unknown, Baltimore*****
Seminole Brand
Ronald L. Newcomb Collection

Back of Above Can

Shelmore Oyster Products Co., Charleston, SC*****
Crystal Bay Brand, 6 ½ Ounces
Carlton and Mary Riggin Collection

L. P. Maggioni & Co., Savannah, GA*****
Daufuski Brand, 11 Ounces
Carlton and Mary Riggin Collection

G. Watson & Co., Savannah, GA*****
Full Moon Brand
Ronald L. Newcomb Collection

Ruge Bro's Canning Co., Apalachicola, FL*****
Alligator Brand
Courtesy of Black Swan Antiques

Back of Above Can

DeJean Packing Co., Biloxi, MS*****
DeJean's Brand, 4 ½ Ounces
Ronald L. Newcomb Collection

DeJean Packing Co., Biloxi, MS*****
Pedigree Brand, 5 Ounces
Courtesy of H. A. Fleckenstein, Jr.

Back of Above Can

Auginbaugh Canning Co., Biloxi, MS*****
Nigger Head Brand, 5 Ounces
Courtesy of Black Swan Antiques

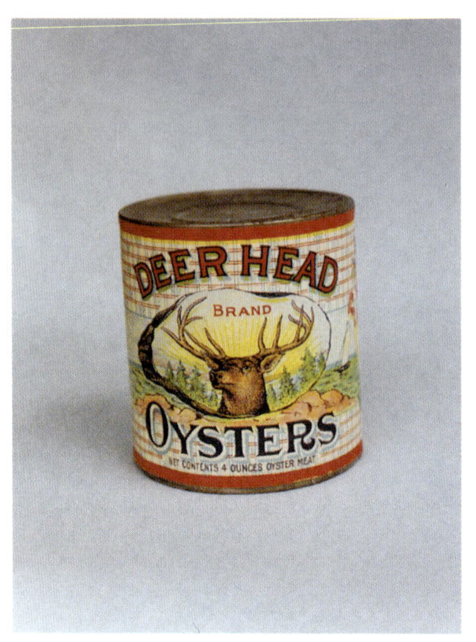

Dunbar-Dukate Co., New Orleans, LA*****
Deer Head Brand, 4 Ounces
Ronald L. Newcomb Collection

Nave-McCord Mercantile Co., St. Joseph, MO****
Frontier Brand, 5 Ounces
Ronald L. Newcomb Collection

Back of Above Can

Dunbar-Dukate Co., New Orleans, LA*****
Pelican Brand, 4 Ounces
Ronald L. Newcomb Collection

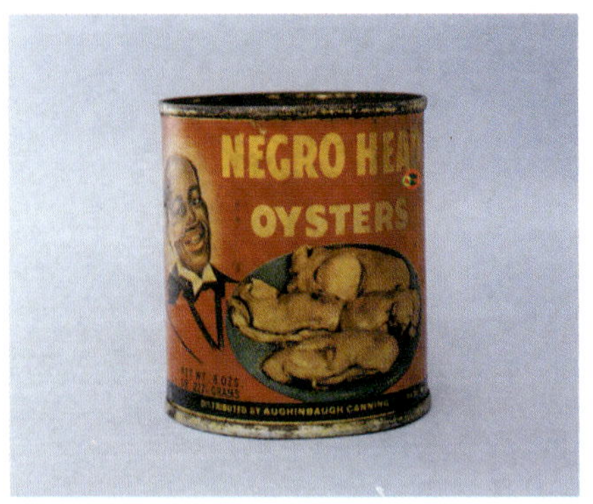

Auginbaugh Canning Co., Biloxi, MS*****
Negro Head Brand, 8 Ounces
Carlton and Mary Riggin Collection

Dorgan McPhillips Pkg. Corp., Mobile, AL****
Sun Set Brand, 4 Ounces

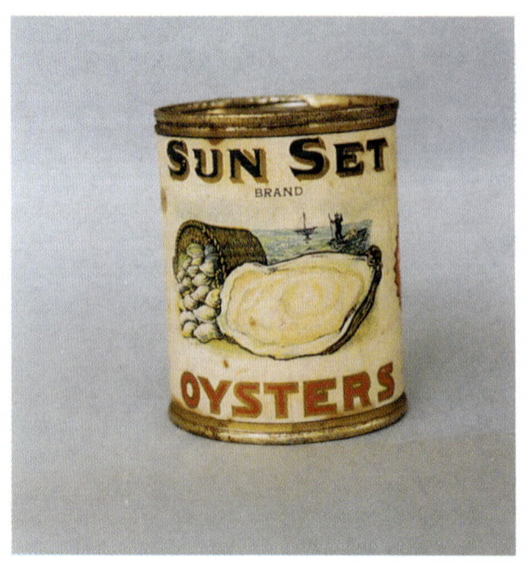

Apple Cupboard Grocery Co., Memphis, TN*****
Gem of the Sea Brand
Carlton and Mary Riggin Collection

Ragland, Potter & Co. of Tennessee*****
Merit Brand, 5 Ounces
Courtesy of H. A. Fleckenstein, Jr.

Hall, Luhrs & Co., Sacramento, CA*****
Rockaway Brand, 5 Ounces
Ronald L. Newcomb Collection

Tillman & Bendel, San Francisco, CA*****
Saddle Rock, 5 Ounces

Royal Blue Stores, Chicago, IL****
Royal Blue Brand, 5 Ounces

Hall, Luhrs & Co. Sacramento, CA*****
Rockaway Brand, 5 Ounces

Wellman-Peck & Co., San Francisco, CA*****
Park Brand, 5 Ounces

Samuel Kunin & Sons, Chicago, IL****
White City Brand, 5 Ounces
Courtesy of Donna Belinko

Mebus & Drescher, Sacramento, CA*****
Mermaid Brand, 6 Ounces
Ronald L.Newcomb Collection

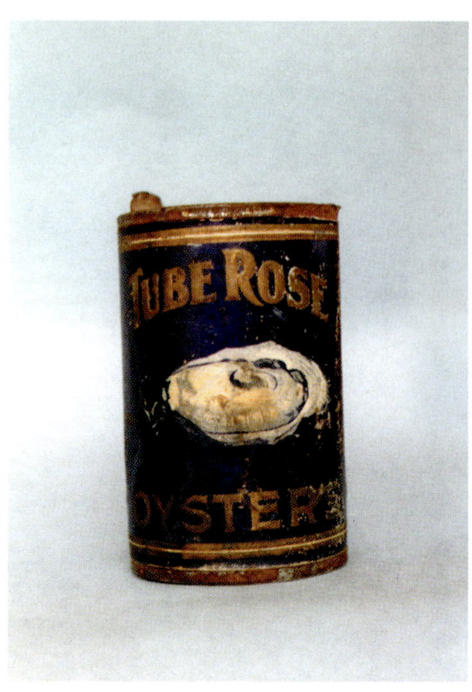

Mebus & Drescher, Sacramento, CA*****
Tube Rose Brand, 5 Ounces

Mebus & Drescher, Sacramento, CA*****
Lady Adams Brand
Ronald L. Newcomb Collection

Back of Above Can

Mebus & Drescher, Sacramento, CA*****
Tube Rose Brand
Ronald L. Newcomb Collection

GLASS BOTTLES

Dorlan & Shaffer, Fulton Market, NY*****
Pickled Oysters
Carlton and Mary Riggin Collection

HO Co., FDL Trademark*****
Oysters on side of Bottle
Ronald L. Newcomb Collection

W.H. McGee & Co., Baltimore, MD*****
Seal Brand, Pint
Licensed by P.F.P. Co., Balt., MD
Patented January 7, 1913
Bill and Steve Dorrell Collection

Smalley, Kivlann Anthank, Boston, MA***
Health Seal Brand Oyster Jar, Quart
Carlton and Mary Riggin Collection

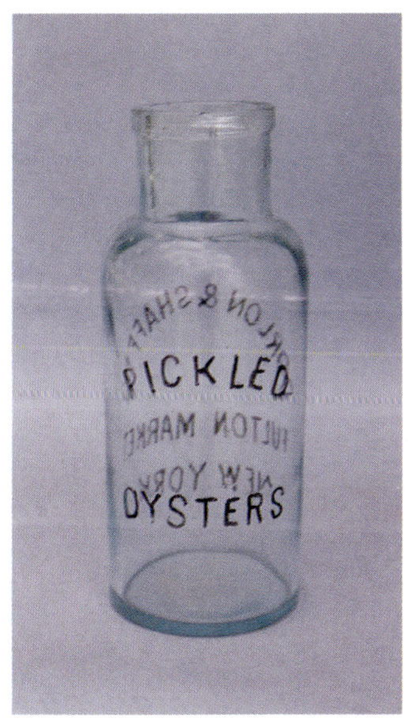

Back of Above Bottle

Rocky Point Oysters****
Finest Quality Brand
Mass. Sealt Bottle Co.
Ronald L. Newcomb Collection

Crisfield Oyster Co., F.H. Market,***
Boston, MA, Quart
Carlton and Mary Riggin Collection

*Wm. Heyser, Baltimore, MD****
Heyser's Oysters, Quart and Pint
Carlton and Mary Riggin Collection*

American Oyster Co., Providence, RI
Note two Different Collars

American Oyster Co., Providence, RI
Oyster Bottle Cap

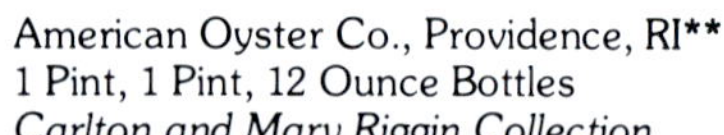

American Oyster Co., Providence, RI**
1 Pint, 1 Pint, 12 Ounce Bottles
Carlton and Mary Riggin Collection

Powers Bros.***
Pint Oyster Bottle
Carlton and Mary Riggin Collection

Oyster Flasks****
Courtesy of H.A. Fleckenstein, Jr.

Piney Island Seafood, Inc., Lancaster, VA*
SEA se we HAK Brand, 12 Ounce and 16 Ounce Jars
Butch and Jackie Cheezum Collection

Fish, Scallop and Two Clam Flasks****
Carlton and Mary Riggin Collection

Food Fair Co., Philadelphia, PA**
Glass Jar

American Oyster Co., Providence, RI**
Saltesea Oysters, From Beds of Unquestioned Purity
8 Ounce and 15 Ounce Jars
Ronald L. Newcomb Collection

The T.A. Snider Preserve Co., Cincinnati, OH***
Two Sniders Oyster Cocktail Bottles
Carlton and Mary Riggin Collection

STONEWARE JUGS

New Haven, CT***
Montauk "Scollops" Bottle
Carlton and Mary Riggin Collection

E.A. Gowen & Son, Dover, NH*****
Stoneware Jug, Gallon
Ronald L. Newcomb Collection

Alfred Jones Sons, Bangor, ME****
Stoneware Jug
Carlton and Mary Riggin Collection

John L. Shriver & Bros., Baltimore, MD****
Shrivers Oyster Ketchup Bottle
Ronald L. Newcomb Collection

G. M. Long & Co., New London CT****
Stoneware Jug
Courtesy of H. A. Fleckenstein, Jr.

Alfred Jones Sons, Bangor Maine****
Stoneware Jug
Courtesy of H.A. Fleckenstein, Jr.

Baltimore, Mch 27th 1865
Mr A. J. Stabler
To ~~Bought of~~ JOHN L. SHRIVER & BROS'
MANUFACTURERS OF
HERMETICALLY SEALED OYSTERS, FRUITS, MEATS, PICKLES, PRESERVES, JELLIES, CATSUPS, &c.
TERMS CASH, IN PAR FUNDS.
SHRIVER'S BALTIMORE OYSTER KETCHUP,
307 West Pratt Street.
Rec'd Payment in full

John L. Shriver & Bros., Baltimore, MD***
Billhead
Ronald L. Newcomb Collection

E. F. Bramhall, Belfast, ME*****.
Stoneware Jug
Courtesy of H.A.Fleckenstein, Jr.

Jones Fish & Oyster House, Bangor, ME*****
Stoneware Jug Quart
Ronald L. Newcomb Collection

Alfred Jones' Sons, Bangor, ME*****
Stoneware Jug
Ronald L. Newcomb Collection

Ackers, "H.G." Registered***
Stoneware Jug
Courtesy of H.A. Fleckenstein, Jr.

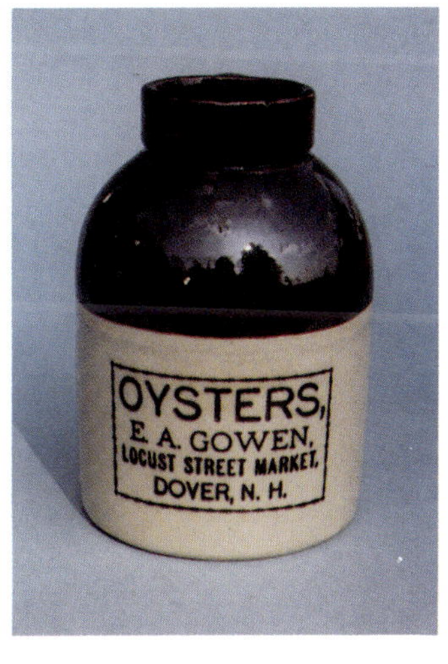

E:A. Gowen, Dover, NH****
Stoneware Jug
Courtesy of H.A. Fleckenstein, Jr.

E. T. Wilson, Farmington, NH****
Stoneware Jug
Carlton and Mary Riggin Collection

Wm. N. Curtis, Medford, MA****
Stoneware Jug
Ronald L. Newcomb Collection

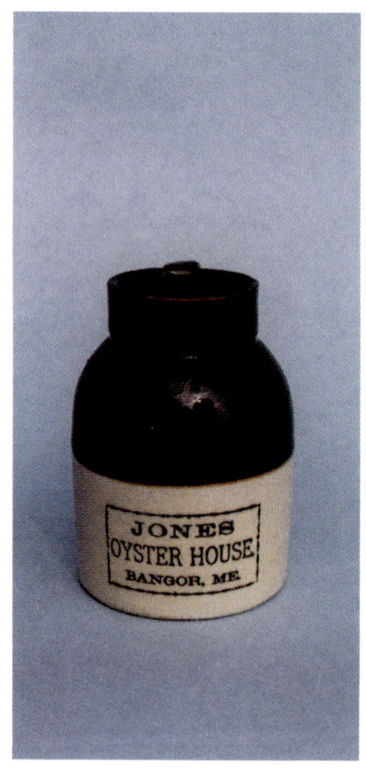

Jones Oyster House, Bangor, ME****
Stoneware Jug
Courtesy of H.A.Fleckenstein, Jr.

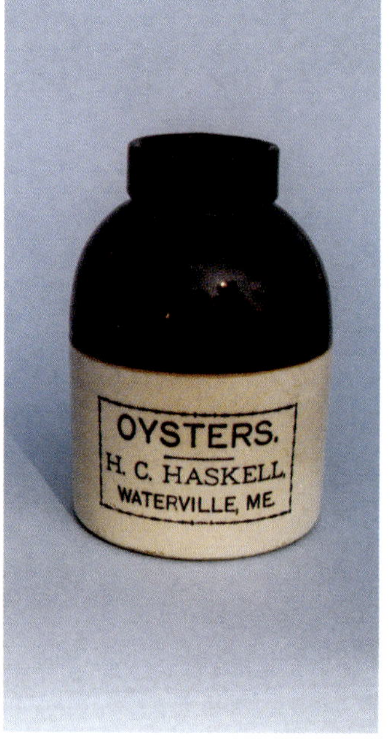

H.C. Haskell, Waterville, ME****
Stoneware Jug
Courtesy of H.A. Fleckenstein, Jr.

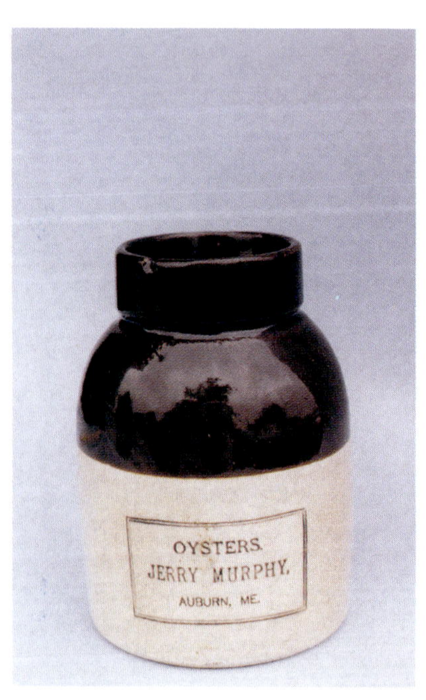

Jerry Murphy, Auburn, ME****
Stoneware Jug
Ronald L. Newcomb Collection

E.A. Bickford, Auburn, ME****
Stoneware Jug
Ronald L. Newcomb Collection

C.C. Porter Fish Co., Bangor, ME****
Stoneware Jug
Courtesy of H.A. Fleckenstein, Jr.

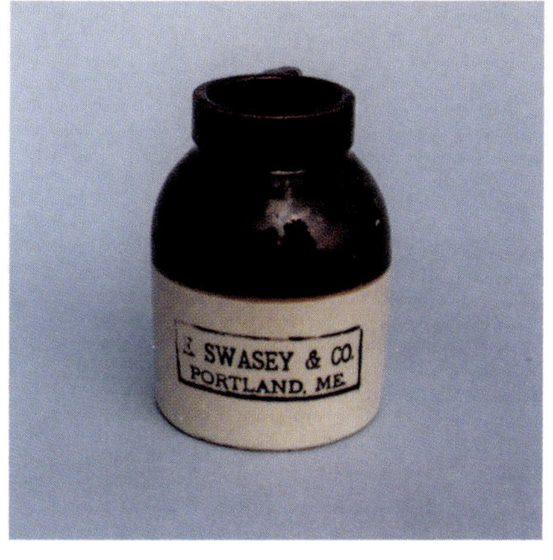

E. Swasey & Co., Portland, ME****
Stoneware Jug
Courtesy of H.A. Fleckenstein, Jr.

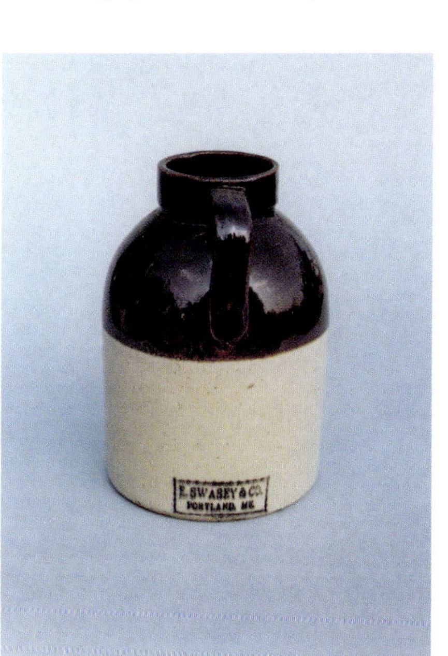

Other Side of Above Jug Showing Handle
Found on Most Jugs

D.H. Seymour, Hartford, CT*****
Two Gallon Stoneware Crock
Courtesy of Black Swan Antiques

John Loveitt & Co., Portland, ME****
Stoneware Jug
Courtesy of H.A. Fleckenstein, Jr.

Cafe Royal Oyster Buffet Crock***
Edinburgh, Scotland
Stoneware Jug
Carlton and Mary Riggin Collection

Thos. Anderson Oysters, Glasgow, Scotland***
Stoneware Jug
Carlton and Mary Riggin Collection

CARDBOARD CARTONS

Hollander & Yeaw, Brattleboro, VT***
Ronald L. Newcomb Collection

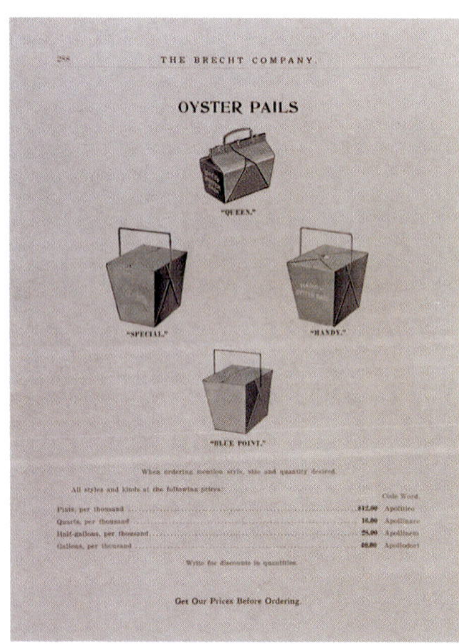

288 THE BRECHT COMPANY.

OYSTER PAILS

"QUEEN."

"SPECIAL." "HANDY."

"BLUE POINT."

When ordering mention style, size and quantity desired.

All styles and kinds at the following prices:

		Code Word.
Pints, per thousand	$12.00	Apolitico
Quarts, per thousand	16.00	Apollinare
Half-gallons, per thousand	28.00	Apollinem
Gallons, per thousand	40.00	Apollodori

Write for discounts in quantities.

Get Our Prices Before Ordering.

The Brecht Co.
Oyster Pail Catalog Page
Joe Sechrist Collection

Blue Points Co., West Sayville, NY**
Sealshipt Brand
Donald C. Bell Collection

Kroger Stores, Cincinnati, OH**
Fres-shore Brand
Courtesy of H.A. Fleckenstein, Jr.

Narragansett Bay Oyster Co., Providence, RI**
Sea Acre Brand
Courtesy of Harris Crab House

W.R. Pittman & Sons, Somers, VA**
Bewdley Brand
Butch and Jackie Cheezum Collection

CRAB MEAT AND OTHER CANS

Crabmeat cans developed at the same time as oyster cans. The earliest had lids tied down with string to ears punched into the sides of the cans. The string was passed under one ear and across the top to the other ear and tied. The color of the string designated the kind of crabmeat in the can: white was claw, green was regular, and red was lump.

Early crabmeat cans were perforated in the bottom to enable water to drain away. In the packing house, shelled crab meat was placed on a board above a tub, the small bits of shell were picked out, and the meat was dumped into the tub which contained water. The meat was drained, iced, and placed into the cans. Melted ice drained out through the holes. After World War II, the state health departments, did away with these holes, citing danger of contamination. Ice was then placed around the cans instead of in them.

Herring roe, shad roe, clam, shrimp and some fish were also packed by a few of these companies. These cans also find their way into collections.

To help in dating containers this short historical sequence may be helpful. Remember no date is absolute, use was phased in or out as each individual company decided to make a change.

1809 Prize for food preservation awarded in France.
1818 Preservation moves to America.
1819 Canning begun in New York.
1839 Tin containers in use.
1849 Canning begun in Baltimore.
1925 Numbered state health certificates issued.
1920s Lithographed cans came into use.
1940s Bail handles began to be phased out.
1963 Zip codes begun.
1970s Plastic containers came into use.

Cedar Creek Packing Co., Wenona, MD*****
Cedar Creek Brand, 1 Pound
Syd Peverley, Jr. Collection

Cedar Creek Packing Co., Wenona, MD*****
Cedar Creek Brand, 1 Pound
Kevin Davidson Collection

G. W. Amory, Jr., Hampton, VA***
Half Pound
Kevin Davidson Collection

Riverside Packing Co., Morgan City, LA***
Riverside Brand, 1 Pound
Kevin Davidson Collection

C.W. Howeth & Bros., Crisfield, MD**
H & B Brand, 1 Pound
Kevin Davidson Collection

Tilghman Packing Co., Tilghman, MD***
Tilghman Brand, 1 Pound
Kevin Davidson Collection

R. W. Strickler, York, PA***
Strickler's, 1 Pound

American Crabmeat Co., Boston, MA**
Sea-Fresh Brand, Half Pound
Kevin Davidson Collection

J.C.W. Tawes & Son, Crisfield, MD***
Tawes Brand, 1 Pound
Kevin Davidson Collection

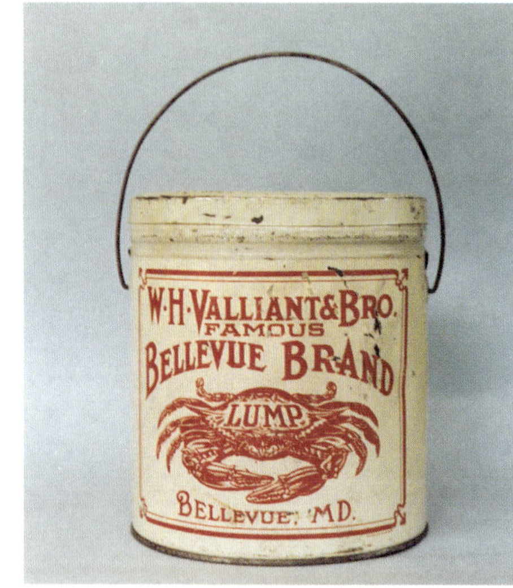

W. H. Valliant & Bro., Bellevue, MD*****
Bellevue Brand, 5 Pounds

Eastern Shore Clam Co., St. Michaels, MD****
Gallon, MD 270
Courtesy of Harris Crab House

I.F. Cannon & Son, Cambridge, MD**
Cannon's Quality Brand, 1 Pound
Kevin Davidson Collection

Same as Above Showing
Tie Down Detail

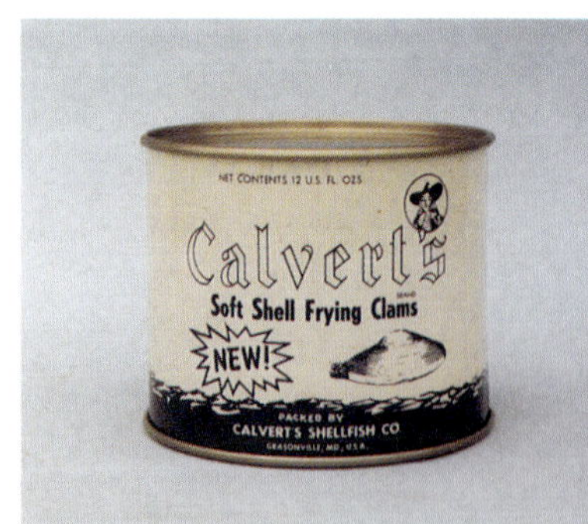

Calvert's Shellfish Co., Grasonville, MD**
Calvert's Brand, 12 Ounces, MD 536
Courtesy of Harris Crab House

Crab Meat Can Display
Kevin Davidson Collection

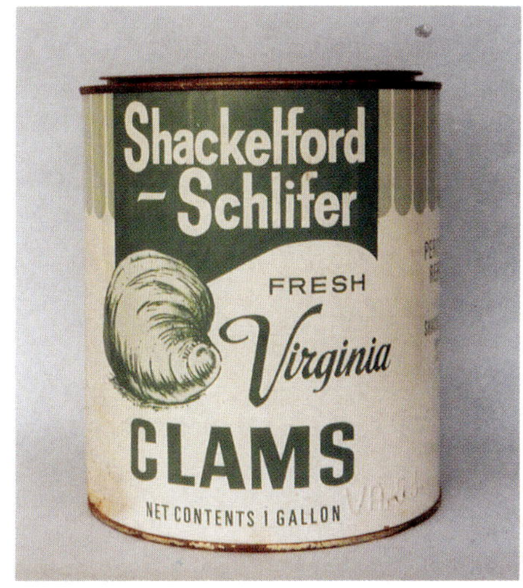

Shackelford-Schlifer Co., Severn, VA**
Shackelford-Schlifer Brand, Gallon, VA 53

Howard Johnson's, Wollaston, MA***
Howard Johnson's Brand, Gallon, MA 150
Ronald L. Newcomb Collection

B & S Fisheries, Inc., Fall River, MA*
B & S Brand, Gallon
Courtesy of Harris Crab House

The R. B. Boak Co., Chicago, IL**
Briny Deep Brand, 10 Pounds
Salt Fish
Ronald L. Newcomb Collection

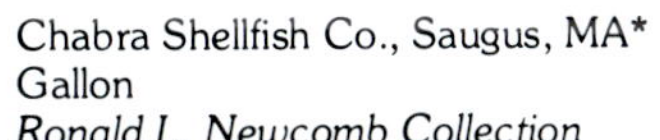

Chabra Shellfish Co., Saugus, MA*
Gallon
Ronald L. Newcomb Collection

General Seafood Co., Halifax, NS****
40-Fathom Brand, Gallon, EXP 366
Donald C. Bell Collection

S & S Cold Storage, Hampton, VA*
8 Pounds Scallops, VA 906

B.A. Griffin Co., Milwaukee, WI*
Dutch Brand, 5 Pounds
Salt Herring
Carlton and Mary Riggin Collection

Ed. Martin Seafood Co., Westwego, LA**
Ed Martin's Brand Shrimp, 5 Pounds
Ronald L. Newcomb Collection

Gulf Central Seafoods, Biloxi, MS**
4 Pounds, Shrimp, MS 106S
Butch and Jackie Cheezum Collection

Vita Food Products, Inc., New York, NY*
5 Pounds, Holland Herring
Carlton and Mary Riggin Collection

Willis Brothers, Williston, NC*
Willis Brand, 8 Pounds Scallops
Ronald L. Newcomb Collection

Russian and Iranian**
Caviar Cans
Carlton and Mary Riggin Collection

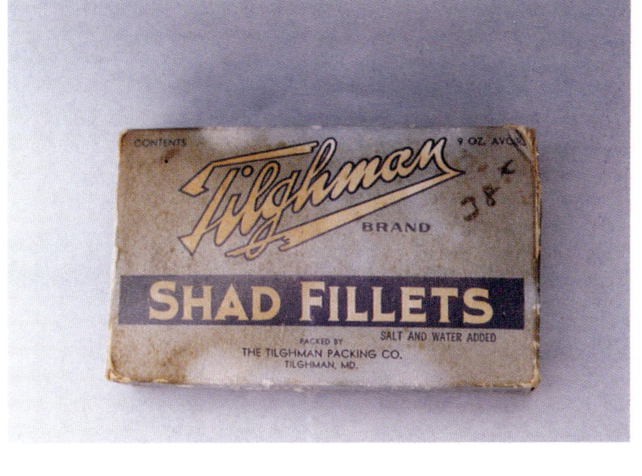

The Tilghman Packing Co., Tilghman, MD***
Tilghman Brand, 9 Ounces
Shad Fillets
David C. Mitchell Collection

Bay City Foods, Baltimore, MD**
Bay City Brand, 5 Pounds Lobster Meat
Mike and Eva Pinder Collection

Woodfield Fish & Oyster Co., Galesville, MD**
Albert W. Woodfield
Woodfield's Brand, Three 1 Pound Cans Herring Roe

Morgan City Packing Co., Houma, LA**
Bayou Rose Brand Shrimp 4 Pounds
Butch and Jackie Cheezum Collection

Chapter II
Oyster Related Items

Collecting oyster plates is a well-established activity, but the same cannot be said of the remainder of oyster related items. The artifacts of the industries that marketed oysters (from the packing or "shucking" houses, as they were called, through the processing, shipping and retailing) are becoming more sought after now. The glamour of the sea contributes to its popularity, and the oyster's connection with nautical, hunting and fishing items gives it charm. Today, collectors are combing the shucking houses, warehouses, antique shops and flea markets where these items may be found. Some oyster items can be purchased, but many others are not for sale; they can only be traded, and old-time traders seem like amateurs compared with oyster traders.

In addition to oyster plates, silverware and cans, other items from the shucking houses, distributors and shippers of oysters are quite interesting. The term "go withs" has been coined to describe some of these things because they are collected to display with collections of decoys, oyster plates and cans, enhancing the overall effect.

Tools the shuckers used such as specialized knives to open oysters and containers to put them in are varied. Company tokens used to keep track of the quantities of oysters shucked were used to pay the workers. Packing house containers, used to move the oysters around in the packing house, along with the specialized boxes and containers used to transport the oysters to the consumers, are varied and interesting. Most of the containers for commercial distribution were designed to accommodate ice while in transit, and some provided a means of dispensing the oysters at the retail store or by peddlers.

The many different papers for keeping track of shipments, invoices, letterheads, envelopes, advertisements and receipt books all are of historical interest today. The fancy engraving used on envelopes and stationery is especially desirable.

SHUCKING KNIVES

Shucking knives are not knives in the usual sense but a tempered blade used to pry open the oyster. Each geographical area had its special technique for opening oysters and each shucker had their own variation. There are two general ways of opening an oyster. In the "stabber" method the knife blade is stabbed or inserted between the two halves of the shell and twisted to sever the muscle. The "cracking" or "billing" method involves placing the end of the oyster on a sharpened iron bar and striking it with the blunt end of the steel crack knife. This action breaks a small part of the edge of the shell and enables the shucker to insert his blade easily to sever the muscle.

Both methods work well and local custom dictated which method was learned and used. A proficient shucker could open a large quantity quickly, since their pay depended on the quantity they produced. Shuckers developed their techniques and modified their knives to their own needs. Many different handle and blade shapes and sizes exist; some even with the shucker's name stamped into the handle to be sure these were not lost. One knife has a three-tined fork forged into the other end, possibly used by the steamed or roasted oyster peddlers to remove these oysters from the partially opened shell or maybe just a convenience to open and eat the oyster with the same tool. A fast worker could shuck twenty gallons a day but fifteen gallons was average, earning the employee $3.00 per day.

There is a wonderful story about a lady living near Charleston, South Carolina that shucked oysters in a unique and prolific way. Observers say she sat with a stake placed between her legs topped by a sharpened piece of metal. She would strike this with an oyster and with one downward movement, split the oyster shell and drop the opened oyster into a pail. She shucked an unbelievable number of oysters this way, sitting on a pile of oyster shells. If she had entered any of the shucking contests, perhaps a new champion would have been crowned.

One very early manufacturer of oyster tools was the John Stortz Company in Philadelphia, founded in 1853 by a German immigrant, experienced in tool making. The company made a full line of oyster shucking tools, the 1902 catalog listing sixteen different styles of knives plus many other related items. These knives are stamped John Stortz, Phila., John Stortz & Son and their economy grade knife was stamped Phila. Tool Co. Today the company, now called John Stortz & Son, continues the business making cast and forged hand tools for many trades. Under the leadership of John C. Stortz, the founders great grandson, it is one of the oldest businesses in continuous operation in Philadelphia.

Another early manufacturer of these knives was the Charles D. Briddell Company of Crisfield, Maryland, making seafood harvesting tools as early as 1895. Mr. Briddell started in a blacksmith shop and built his business into one of the countries largest suppliers of these tools. In one catalog eighteen different styles of oyster shucking knives were listed, covering regional variations from New England to Texas. An equally varied line of crab knives and oyster tongs were available. In an attempt to make shucking more efficient, the company developed a grinder to remove the part of the shell that the crack knife broke off. The shucker held the shell against the grinding wheel to take off the edge so the knife could be inserted. This method did not catch on and soon passed out of existence. Today the company continues to make these knives under the new company name of Carvel Hall. An interesting collection could be made by finding one of each of these different styles.

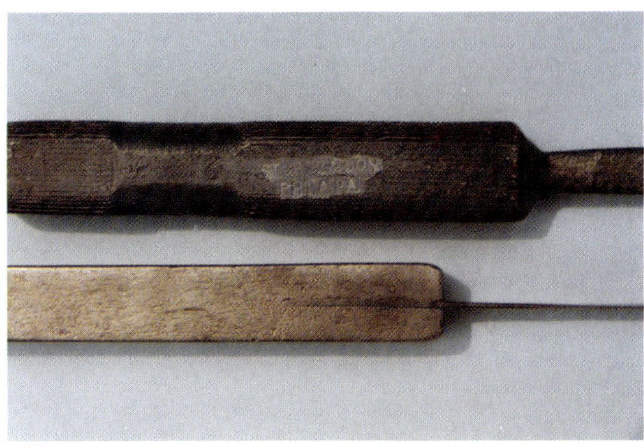

Stortz Crack Knife**
Knife with Blade Inserted in Handle

Shucking Block with Crack Knife**
Ralph and Betty Tull Collection

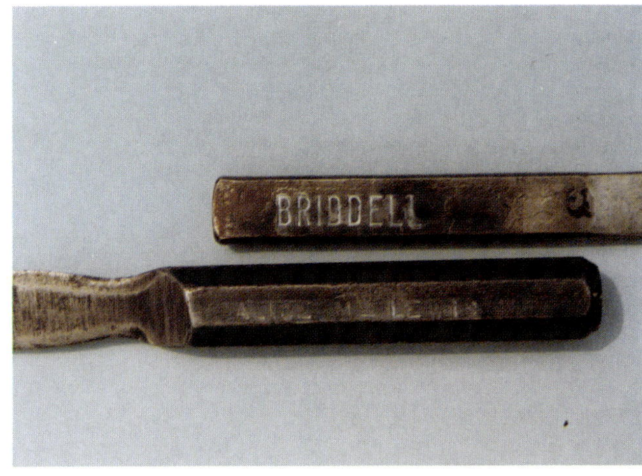

Briddell Knife and Personalized Shuckers Knife**

Sizes of Stortz Crack Knives**

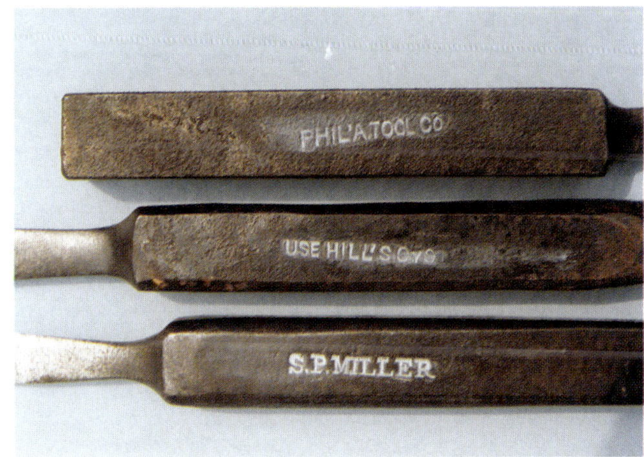

Three Crack Knives**

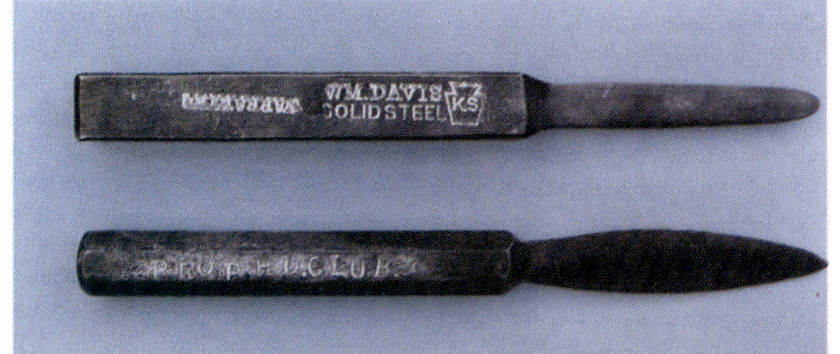

Two Crack Knives**

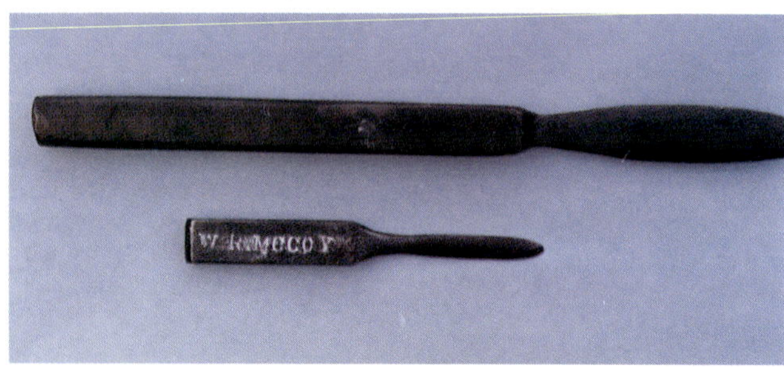

Crack Knives 11-¼ and 5-¼ Inches**
Longest and Shortest

Crack Knife with Fork***
Crack Knife with Folding Fork***
Courtesy of Harris Crab House

Three Crack Knives**
Roberta and Rudy Schmehl Collection

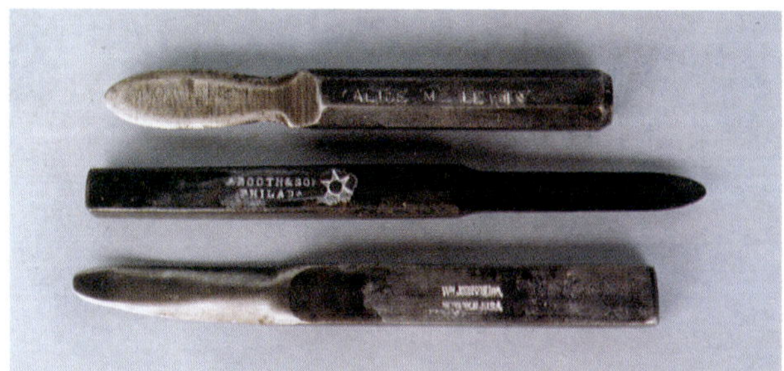

Three Crack Knives**

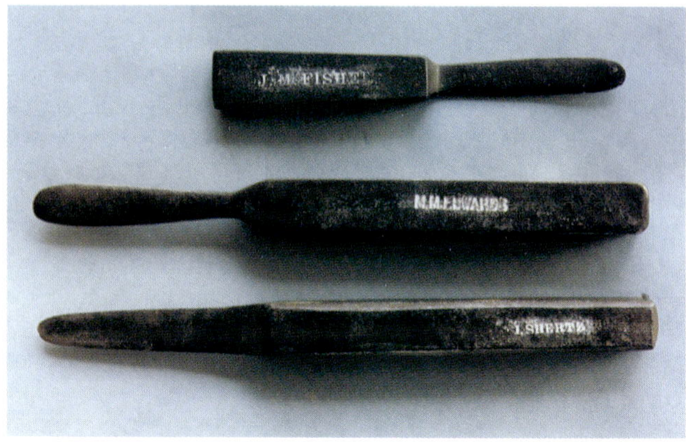

Three Crack Knives made by***
Lancaster Co., PA Blacksmiths

Four Crack Knives, One with Fork***
Courtesy of Joe H. Thome

Three Stabbing Knives**

Briddell Stabbing Knife**

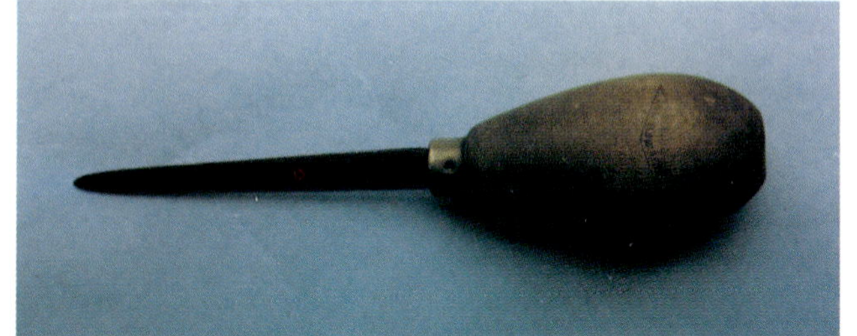

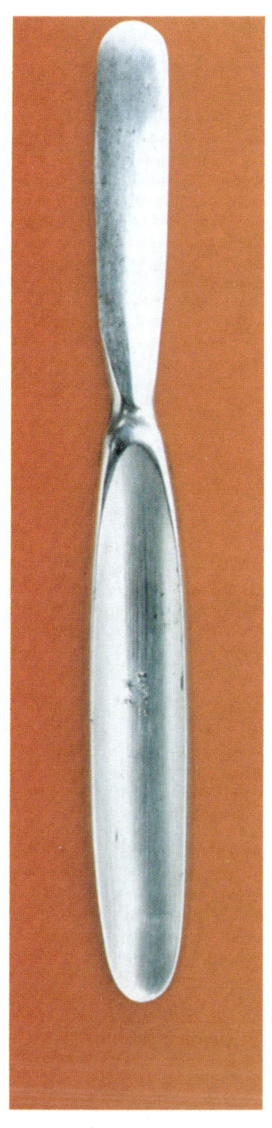

Shucking knife made by Landers, Frary & Clark in 1918****

Detail of shucking knife

Winchester Stabbing Knife***

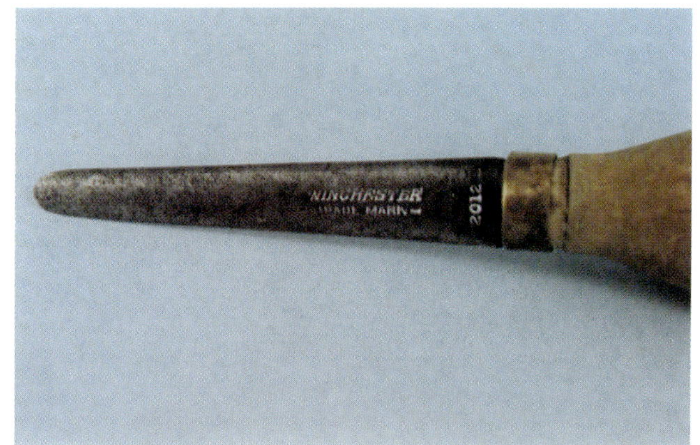

Detail of Above

Two Advertising Stabbing Knives**

3 Wooden Handle Knives**

Briddell Clam Knife**

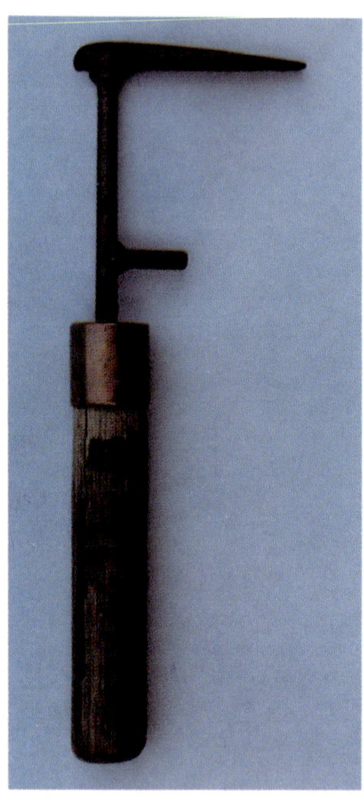

Culling Knife with Three Inch Measuring Guage**

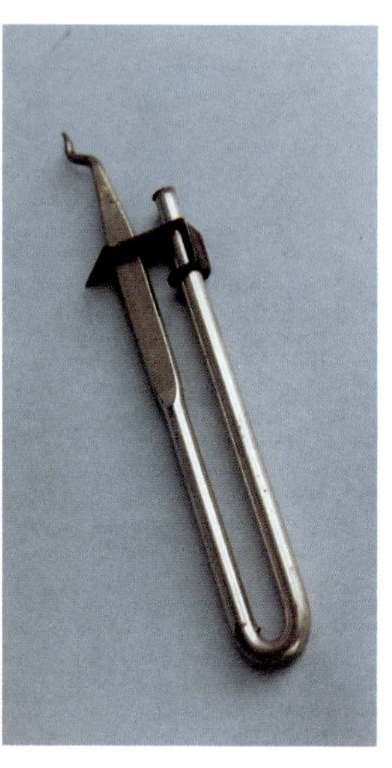

Can Opener For Plug Top Cans***
Ronald L. Newcomb Collection

Maryland Conservation Department***
Measuring Guage, Three Inch Requirement

John Stortz & Son, Philadelphia, PA***
1902 Catalog

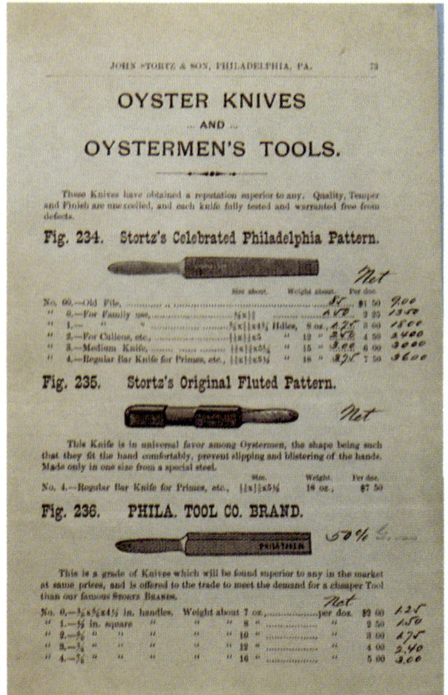

JOHN STORTZ & SON, PHILADELPHIA, PA. 73

OYSTER KNIVES
AND
OYSTERMEN'S TOOLS.

These Knives have obtained a reputation superior to any. Quality, Temper and Finish are unexcelled, and each knife fully tested and warranted free from defects.

Fig. 234. Stortz's Celebrated Philadelphia Pattern.

	Size about.	Weight about.	Per doz.
No. 00.—Old File,			$1 50
" 0.—For Family use,	½x1½		2 25
" 1.— " "	⅝x1½x4½ Hdles,	8 oz.,	3 00
" 2.—For Cullens, etc.,	[illegible]	" 12 "	4 50
" 3.—Medium Knife,	[illegible]	" 15 "	6 00
" 4.—Regular Bar Knife for Primes, etc.,	[illegible]	" 16 "	7 50

Fig. 235. Stortz's Original Fluted Pattern.

This Knife is in universal favor among Oystermen, the shape being such that they fit the hand comfortably, prevent slipping and blistering of the hands. Made only in one size from a special steel.

	Size.	Weight.	Per doz.
No. 4.—Regular Bar Knife for Primes, etc.,	[illegible]	16 oz.,	$7 50

Fig. 236. PHILA. TOOL CO. BRAND.

This is a grade of Knives which will be found superior to any in the market at same prices, and is offered to the trade to meet the demand for a cheaper Tool than our famous STORTZ BRANDS.

No. 0,—[illegible] in. handles, Weight about 7 oz.,	per doz.	$2 00
" 1.—½ in. square " " " 8 "	"	2 50
" 2.—[illegible] " " " " 10 "	"	3 00
" 3.—[illegible] " " " " 12 "	"	4 00
" 4.—[illegible] " " " " 16 "	"	5 00

J Stortz & Son, Philadelphia, PA
Catalog Pages showing detail and prices of Steel Crack Knives

80 JOHN STORTZ & SON, PHILADELPHIA, PA.

Fig. 269. CULLING IRONS.

	Handled.	Heads only
No. 5.—Malleable Iron; Bell Head; Style of Cut,....per doz.,	$1 20	$0 75

Fig. 260. OYSTER HARDIES.

Fig. 261.

Style of Nos. 5 and 6. Style of No. 4.

	Height	Price, each,
No. 1.—Solid Triangle, light,	7½ in.	$0 75
" 2.—Block Hardie,	6 "	75
" 3.—Hollow Triangle,	7½ "	95
" 4.—Solid Triangle with Ring, heavy,	7½ "	1 00
" 5.—Forged Post Hardies,	6 to 12 "	1 05
" 6.— " " "	above 12 in.,	1 15

Fig. 262. NEW YORK PATTERN.

No. 7.—Lignum Vitæ Blocks; Steel Posts,..........Price, each, $1 15
" 8.—Breaks only,.......... " per doz., 1 00

J. Stortz & Son, Philadelphia, PA
Catalog Pages Of Oyster Shucking Tools

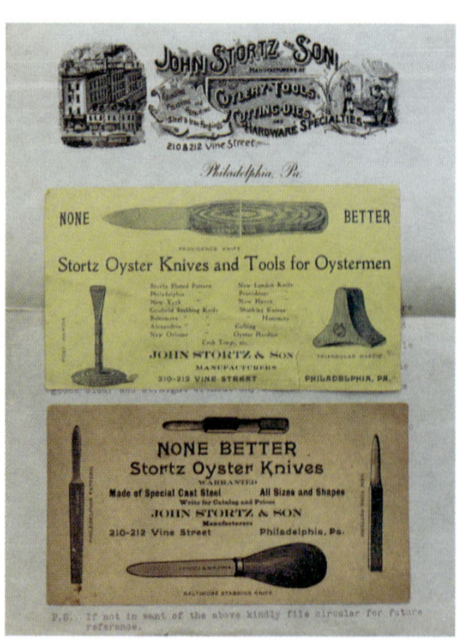

J. Stortz & Son, Philadelphia, PA***
Advertising and Letterhead

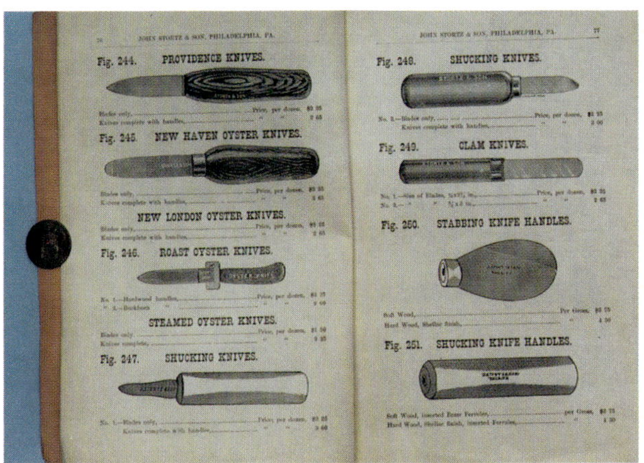
JOHN STORTZ & SON, PHILADELPHIA, PA.

Fig. 244. PROVIDENCE KNIVES.

Fig. 245. NEW HAVEN OYSTER KNIVES.

NEW LONDON OYSTER KNIVES.

Fig. 246. ROAST OYSTER KNIVES.

STEAMED OYSTER KNIVES.

Fig. 247. SHUCKING KNIVES.

JOHN STORTZ & SON, PHILADELPHIA, PA. 77

Fig. 248. SHUCKING KNIVES.

Fig. 249. CLAM KNIVES.

Fig. 250. STABBING KNIFE HANDLES.

Fig. 251. SHUCKING KNIFE HANDLES.

John Stortz & Son, Philadelphia, PA
Catalog Pages showing detail and prices of
Wooden Handled Oyster Knives

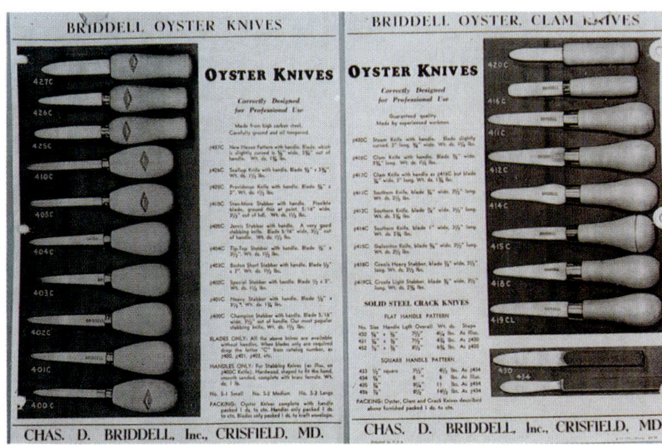
BRIDDELL OYSTER KNIVES

OYSTER KNIVES

Correctly Designed
for Professional Use

CHAS. D. BRIDDELL, Inc., CRISFIELD, MD.

BRIDDELL OYSTER, CLAM KNIVES

OYSTER KNIVES

Correctly Designed
for Professional Use

SOLID STEEL CRACK KNIVES

FLAT HANDLE PATTERN

SQUARE HANDLE PATTERN

CHAS. D. BRIDDELL, Inc., CRISFIELD, MD.

Chas. D. Briddell Co., Crisfield, MD**
Catalog Pages Showing Line of Oyster Knives
Courtesy of Carvel Hall

SHUCKING TOKENS

In some packing houses tokens were given to shuckers when a pre-determined quantity of oysters was opened and they are a large and varied lot. Tokens usually were made from metal because of the wet conditions and were stamped with designs and the quantity (one gallon, five gallons, etc.). The packer's name was often used, but some simply had a design like a masonic emblem, an oyster shell, or anything else distinctive to that packer. At the turn of the century the usual pay for shucking a gallon of oysters was twenty cents, a number often included on the token.

Platt & Co., Baltimore, MD***
Thomas Kensett & Co., Baltimore, MD***
Russ Sears Collection

Lord Mott Co., Baltimore, MD***
William H. Killian & Co., Baltimore, MD***
Russ Sears Collection

W.H. Valliant & Bro., Bellevue, MD***
William H. Killian & Co., Baltimore, MD***
Russ Sears Collection

L.T. Dryden & Co., Crisfield, MD***
T. D. Twaits & Co., Baltimore, MD***
Russ Sears Collection

L. B. Phillips & Co., Cambridge, MD***
Good Luck Penny
Russ Sears Collection

Thomas J. Myers & Co., Baltimore, MD***
S.S. Barnes & Co., Baltimore, MD***
Russ Sears Collection

Brown & Bennett, Deal Island, MD ***
Bill and Steve Dorrell Collection

T.H.C. & Co.,***
Chesapeake Oyster Co.***
Russ Sears Collection

Reverse of the Above
Russ Sears Collection

Price Bros. & Co., Baltimore, MD (Bros. Overstruck)***
Russ Sears Collection

Wm. L. Ellis & Co., Baltimore, MD***
Reverse Side Price, Mann & Co., Balt. MD
Russ Sears Collection

Price Mann & Co., Baltimore, MD***
Reverse Side Wm. L. Ellis & Co., Balt. MD
Russ Sears Collection

Reverse of Above

O'Neill & Co., Canton, Baltimore, MD***
L.C. Spencer & Co. Baltimore, MD***
Russ Sears Collection

Two Leatherbury Bros. & Co., Shady Side, MD**
Costen & Blades, S.S. Costen & Co.,***
Tull & Co., Kingston, MD

REVENUE STAMPS

Another collectible item is the revenue stamps issued by South Carolina and Georgia taxing raw shucked and canned oysters. These stamps were affixed to either the can or case of oysters when they were shipped from the packer.

South Carolina first issued these stamps in 1906 and Georgia in 1925. South Carolina discontinued them in the 1950's or possibly in early 1960's while Georgia ended the tax in 1953. David R. Torre, a specialist in wildfowl stamps, says there were quite a number of different designs, a few are shown here.

Georgia Revenue Stamps****
Courtesy of David R. Torre

South Carolina Revenue Stamps****
Courtesy of David R. Torre

SHIPPING AND STORE CONTAINERS

Oysters were shipped in wooden crates, starting with the canned oysters sent over the country by wagon, steamboat and train. Colorful graphics were printed on the wood, creating a large field for collectors. Later when shucked oysters were shipped in cans they were also placed in wooden boxes surrounded by ice, renewed during transit. Some companies shipped their oysters in containers not only suitable for shipment, but also suitable for display in the retail store. The Sealshipt Company was a leader in this, but other companies and individual packers also used them. A display of these containers is used as a background by some can collectors.

T.J. Myer & Co., Baltimore, MD***
Maryland Brand
Carlton and Mary Riggin Collection

Miller, Bros. & Co., Baltimore, MD****
Jumbo Brand
Ronald L. Newcomb Collection

C.H. Pearson & Co., Baltimore, MD***
Peerless Brand
Ronald L. Newcomb Collection

Martin Wagner Co., Baltimore, MD***
Dog's Head Brand
Ronald L. Newcomb Collection

C.H. Lighthiser, Baltimore, MD***
Elephant Brand
George and Betty Juergens Collection

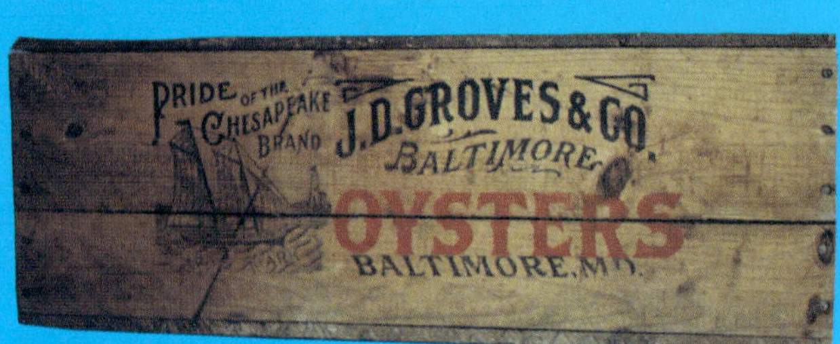

J.D. Groves & Co., Baltimore, MD***
Pride of the Chesapeake Brand
Carlton and Mary Riggin Collection

Gibbs Preserving Co., Baltimore, MD***
Bulls Head Brand
Carlton and Mary Riggin Collection

Webster-Butterfield Co., Baltimore, MD***
Webster's Best Brand
Ronald L. Newcomb Collection

Wm. Fait Co., Baltimore, MD***
Monkey Brand
Ronald L. Newcomb Collection

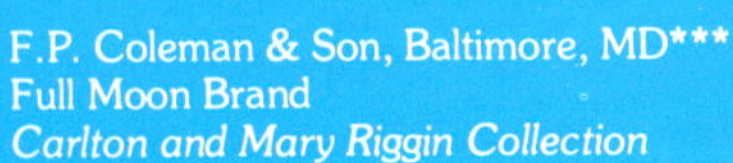

F.P. Coleman & Son, Baltimore, MD***
Full Moon Brand
Carlton and Mary Riggin Collection

Louis Grebb & Co., Baltimore, MD***
Belle Brand
Carlton and Mary Riggin Collection

R. E. Roberts Co., Baltimore, MD***
Maryland Beauty Brand
Ronald L. Newcomb Collection

Southern Seafood Co., Baltimore, MD**
Bill and Steve Dorrell Collection

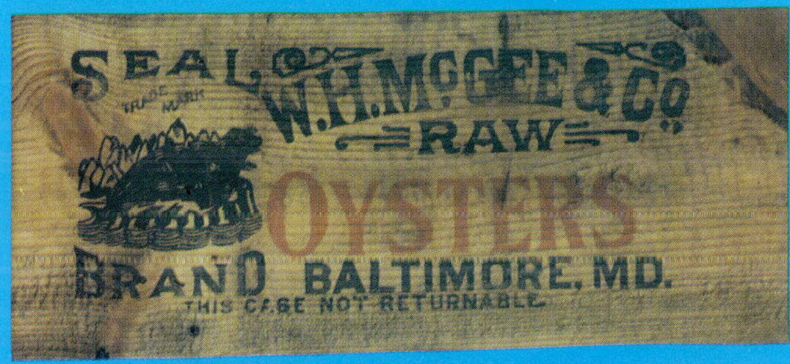

W.H. McGee & Co., Baltimore, MD***
Seal Brand
Syd Peverley Jr. Collection

Coulbourn Bros. Co., Baltimore, MD***
Sea C Brand
Ronald L. Newcomb Collection

Caldwell Bros. & Co., Baltimore, MD***
Monitor Brand
Ronald L. Newcomb Collection

J. B. Brinkley & Sons, Baltimore, MD***
J. B. & Sons Brand
Ronald L. Newcomb Collection

Wm. D. Gude & Co., Baltimore, MD***
Premium Brand
Ronald L. Newcomb Collection

Planters Trading Co., Baltimore, MD***
Stag Brand
George and Betty Juergens Collection

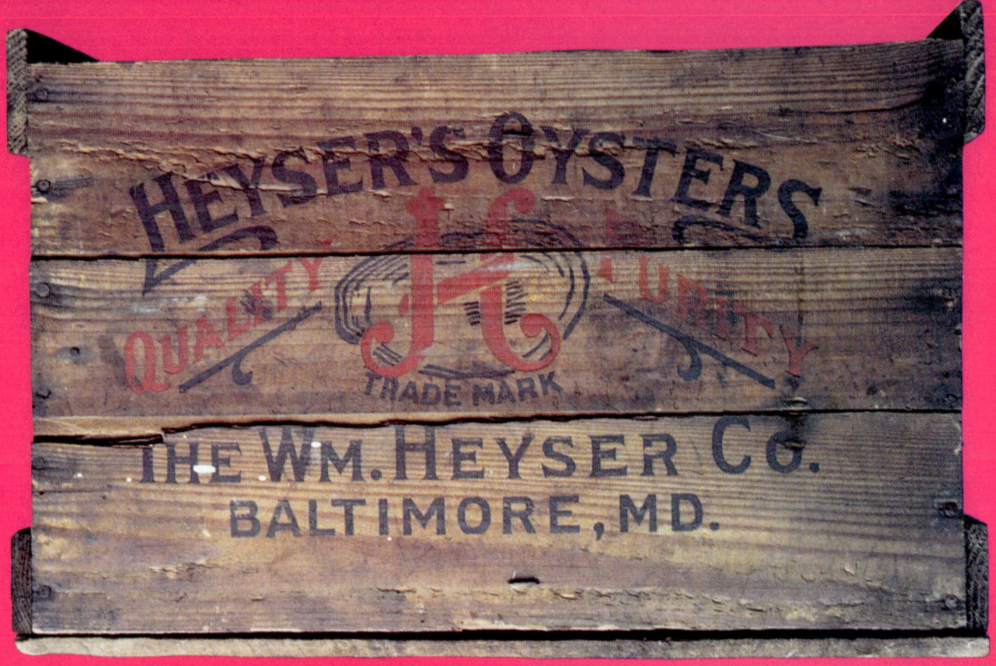

Wm. Heyser Co., Baltimore, MD***
Heyser's Brand
Ronald L. Newcomb Collection

The Leib Packing Co., Baltimore, MD***
Sun Brand

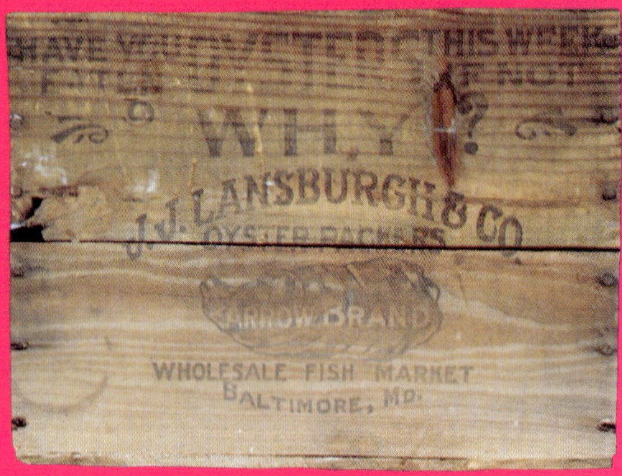

J. J. Lansburgh & Co., Baltimore, MD***
Arrow Brand
"Have You Eaten Oysters This Week,
If Not Why?"
Ronald L. Newcomb Collection

Shelter Island Oyster Co., Greenport, NY**
Shelter Island Brand
Butch and Jackie Cheezum Collection

John T. McNaney, Baltimore, MD***
McNaney Brand
Ronald L. Newcomb Collection

Oyster Bay Oyster Co., Oyster Bay, NY***
Seawanhaka Brand
Ronald L. Newcomb Collection

Thurston & Kingsbury, Bangor, ME***
T & K Brand
Carlton and Mary Riggin Collection

J. & J. W. Elsworth Co., Greenport, NY***
"Fireplace Oysters"
Carlton and Mary Riggin Collection

Olympia Oyster Co., Oyster Bay, WA**
Clams
Butch and Jackie Cheezum Collection

Shelter Island Oyster Co., Greenport, NY**
Shelter Island Brand
Carlton and Mary Riggin Collection

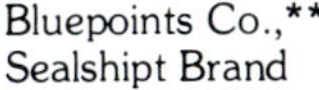

Bluepoints Co.,**
Sealshipt Brand

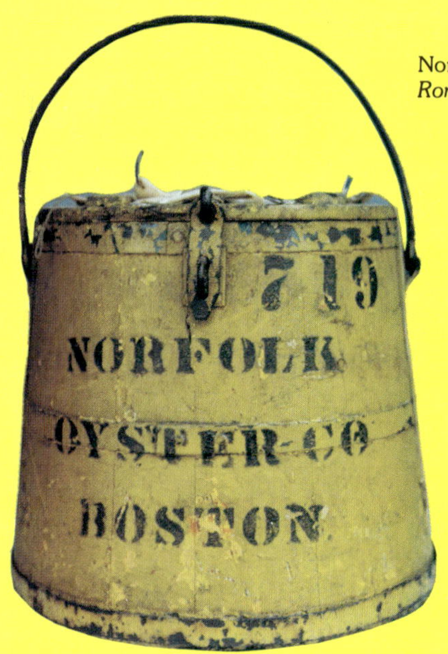

Norfolk Oyster Co., Boston, MA****
Ronald L. Newcomb Collection

C. Harrop, Baltimore, MD****
Carlton and Mary Riggin Collection

Narragansett Bay Oyster Co., Providence, RI**
Ronald L. Newcomb Collection

Shipping Label for a Barrel
Carlton and Mary Riggin Collection

Booth Fisheries Corp., Chicago IL**
Booth Brand

R.E. Roberts Co., Baltimore, MD****
Maryland Beauty Brand, 5 Gallon
Ronald L. Newcomb Collection

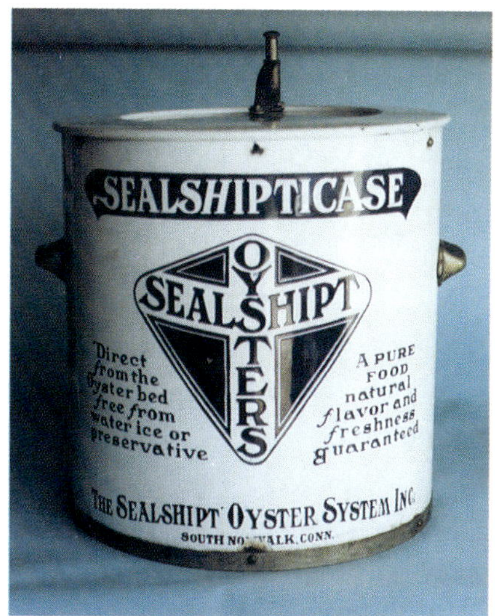

The Sealshipt Oyster System Inc, South Norwalk, CT***
Sealshipt Shipping Case
Carlton and Mary Riggin Collection

Five Gallon Graduated Measuring Can**
Carlton and Mary Riggin Collection

Neubert & Co, Baltimore, MD**
Five Gallon Stainless Steel Shucking House Bucket
Carlton and Mary Riggin Collection

Standard Fish & Oyster Co, Apalachicola, Florida****
Ronald L. Newcomb Collection

H. McWilliams & Co., Baltimore, MD****
Retail Oyster Display
Ronald L. Newcomb Collection

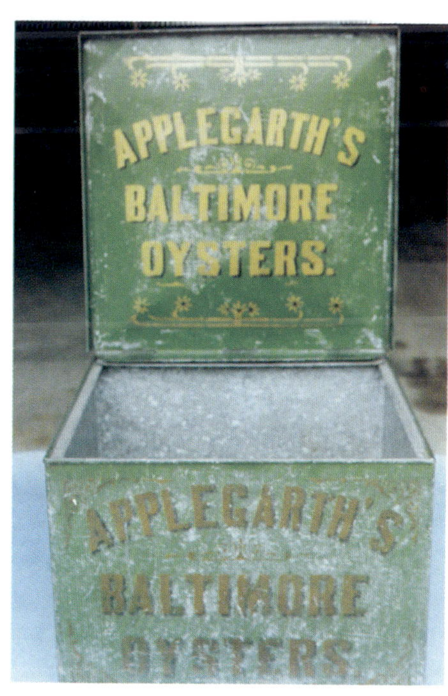

C. L. Applegarth & Co.****
Retail Oyster Display
Ronald L. Newcomb Collection

Oyster Shipping Container***
Carlton and Mary Riggin Collection

Three Gallon Shucking House Buckets**
Carlton and Mary Riggin Collection

Shucking House Draining Pan*
Carlton and Mary Riggin Collection

LABELS, ADVERTISEMENTS, AND STATIONERY

Oyster companies advertised just as any other company did. Colorful posters, signs and give-aways were used. Some of the signs show great artwork and color. Companies, in designing their advertising, developed logos and phrases they wanted to protect from use by competitors. The copyright laws accomplish this and the federal government is responsible for enforcing them. The designs are published in the same way as design patents are and can be found by searching the patent office files. The fancy style of printing popular in the past used steel engravings of brand names, ships, buildings and lettering. Stationery and other forms of printed paper memorabilia can be placed into albums or framed, and should be treated carefully like other historical papers.

The John Boyle Co., Baltimore, MD***
Victory Brand Can Label
Joe Sechrist Collection

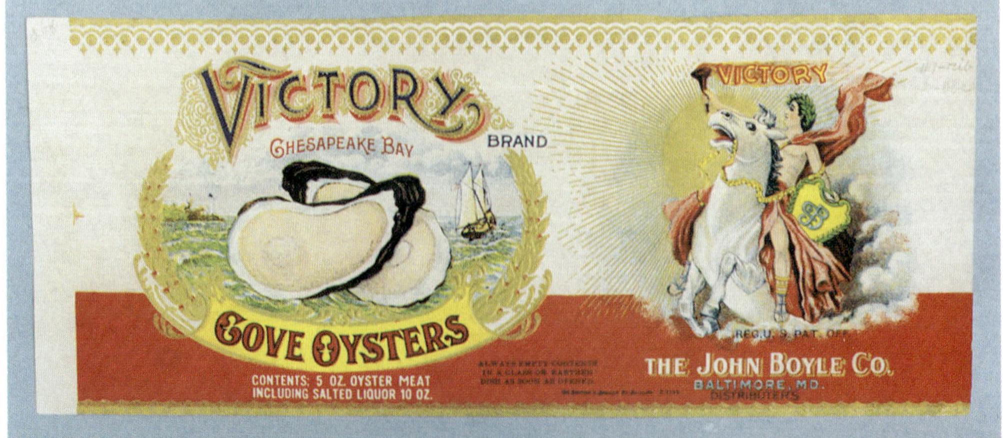

F. A. Waidner & Co., Baltimore, MD****
Can Label

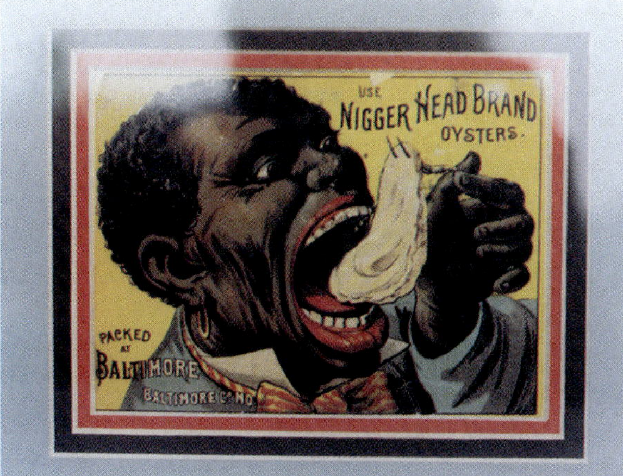

Auginbaugh Canning Co., Baltimore, MD*****
Nigger Head Brand Can Label
Ronald L. Newcomb Collection

Can Labels***
Dunbar-Dukate Co., New Orleans. LA
Deer Head Brand;
Carnaveron Canning Co., Carnaveron, LA
Pride of Gulf Brand;
Southern Shell Fish Co., Harvey, LA
Palm Brand
Joe Sechrist Collection

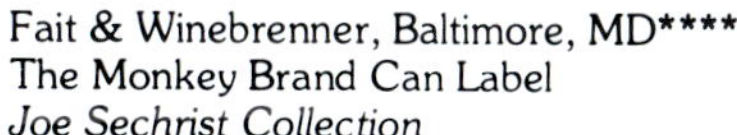

Fait & Winebrenner, Baltimore, MD****
The Monkey Brand Can Label
Joe Sechrist Collection

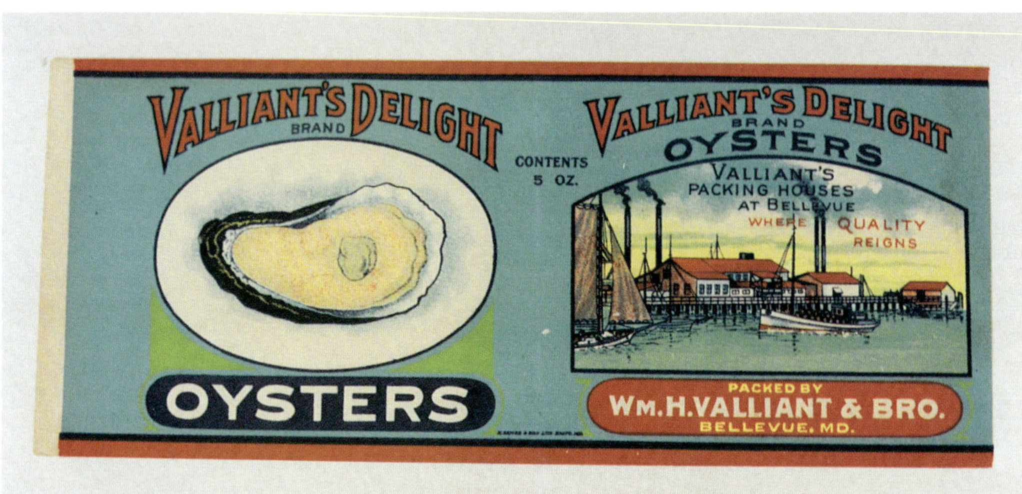

Wm. H. Valliant & Bro., Bellevue, MD***
Valliant's Delight Brand Can Label

Annapolis Canning Co., Annapolis, MD****
Midshipman Brand Can Label
Ronald L. Newcomb Collection

John Hartman, Philadelphia, PA***
"Blue Points" Oyster Crackers Trade Card
Note Crackers Shaped Like Oysters

Can Labels***
Martin Wagner Co., Baltimore, MD
Cherry-Stone Lunch Oysters;
Englehard Shrimp Fish & Oyster Co.,
Englehard, NC Chief Englehard;
Moore & Brady, Baltimore, MD
Deep Sea Brand

George Taulane & Co.,
Philadelphia, PA**
Envelope and Shipping Tag
"Salt and Fresh Oysters"

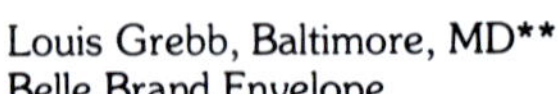

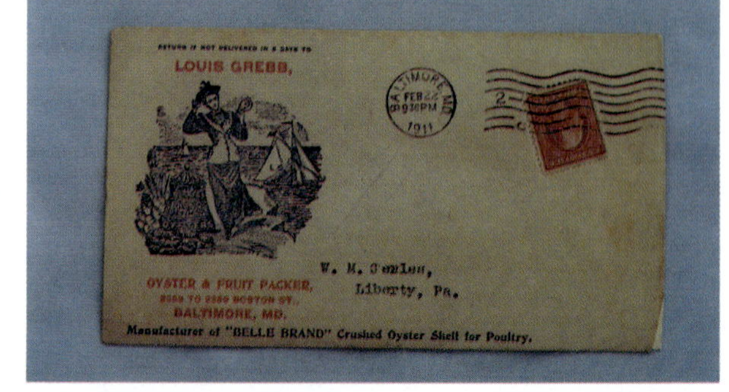

Louis Grebb, Baltimore, MD**
Belle Brand Envelope

Wm. H. Valliant & Bros., Bellevue, MD**
Bellevue Brand Envelope
Joe Sechrist Collection

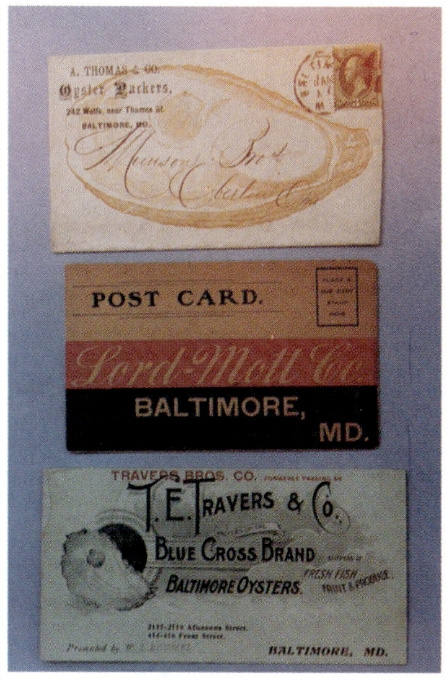

A. Thomas & Co., Baltimore, MD**
Lord Mott Co., Baltimore, MD
T.E. Travers & Co., Baltimore, MD
Blue Cross Brand, Envelopes and Post Card

C.H. Pearson Packing Co., Baltimore, MD**
Peerless Brand Envelope

Wm. A. Keagle & Son, Baltimore, MD
Pimlico Brand**
A. Phillips & Co., Cambridge, MD
J.H. Collison, Baltimore, MD Xc'lent Brand Envelopes

William Heyser, Baltimore, MD, Billhead**
Wm. B. McCaddin & Co., Baltimore, MD Irma Brand Envelope

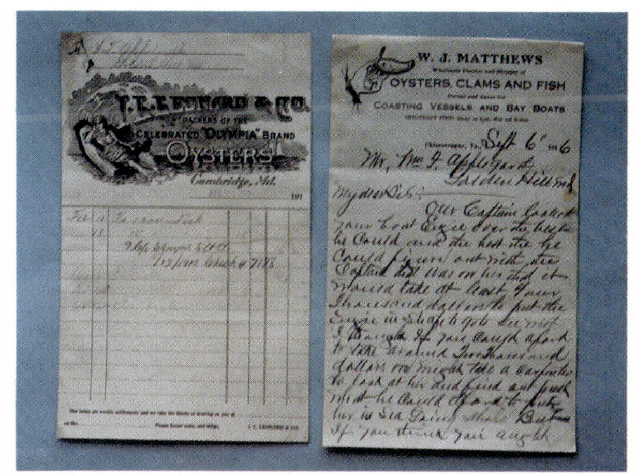

I.L. Leonard & Co., Cambridge, MD**
Olympia Brand Billhead;
W.J. Matthews, Chincoteague, VA Letter

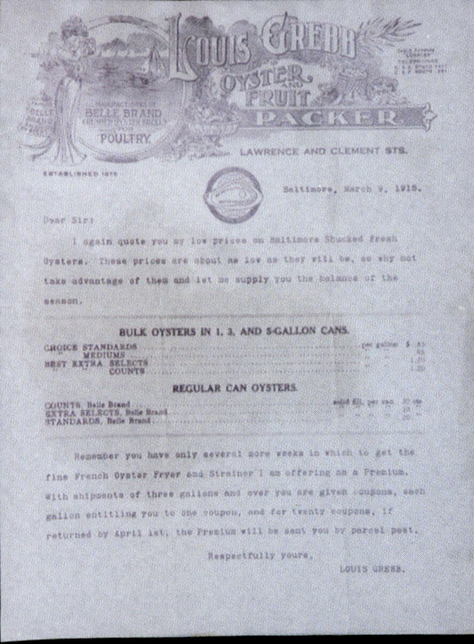

LOUIS GREBB
OYSTER AND FRUIT PACKER
BELLE BRAND
POULTRY
LAWRENCE AND CLEMENT STS.
ESTABLISHED 1875

Baltimore, March 9, 1915.

Dear Sir:

I again quote you my low prices on Baltimore Shucked Fresh Oysters. These prices are about as low as they will be, so why not take advantage of them and let me supply you the balance of the season.

BULK OYSTERS IN 1, 3, AND 5-GALLON CANS.

CHOICE STANDARDS
MEDIUMS
BEST EXTRA SELECTS
COUNTS

REGULAR CAN OYSTERS.

COUNTS, Belle Brand
EXTRA SELECTS, Belle Brand
STANDARDS, Belle Brand

Remember you have only several more weeks in which to get the fine French Oyster Fryer and Strainer I am offering as a Premium. With shipments of three gallons and over you are given coupons, each gallon entitling you to one coupon, and for twenty coupons, if returned by April 1st, the Premium will be sent you by parcel post.

Respectfully yours,
LOUIS GREBB.

Louis Grebb, Baltimore, MD**
Letter with Price Quotes

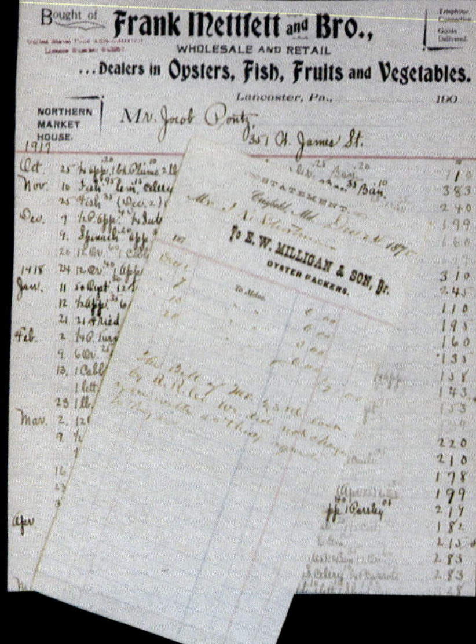

Bought of Frank Mettfett and Bro.,
WHOLESALE AND RETAIL
...Dealers in Oysters, Fish, Fruits and Vegetables.
Lancaster, Pa.
NORTHERN MARKET HOUSE.

STATEMENT.
Crisfield, Md.
To E. W. MILLIGAN & SON, Dr.
OYSTER PACKERS.

Frank Mettfett & Bro., Lancaster, PA**
E.W. Milligan & Son, Crisfield, MD
Billheads

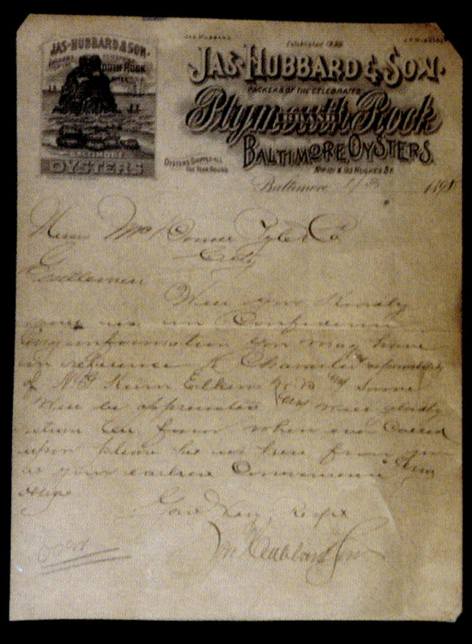

JAS. HUBBARD & SON.
PACKERS OF THE CELEBRATED
Plymouth Rock
BALTIMORE OYSTERS
Baltimore

Jas. Hubbard & Sons, Baltimore, MD**
Plymouth Rock Brand, Letter
Courtesy of Harris Crab House

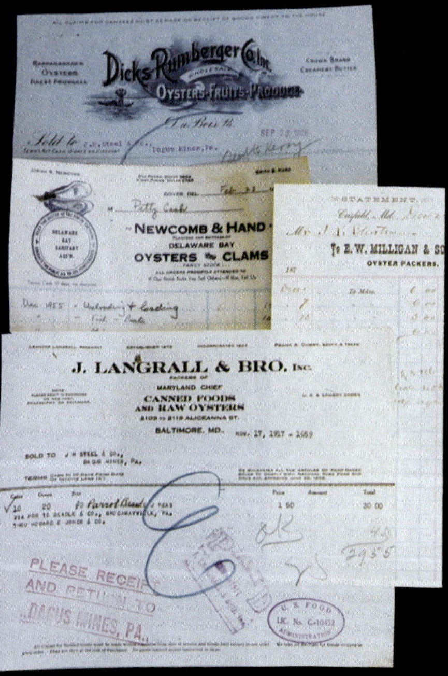

Dicks-Rumberger Co. Inc.
OYSTERS-FRUITS-PRODUCE
Du Bois, Pa.

NEWCOMB & HAND
DELAWARE BAY
OYSTERS AND CLAMS

STATEMENT.
To E. W. MILLIGAN & SON
OYSTER PACKERS.

J. LANGRALL & BRO. INC.
PACKERS OF
MARYLAND CHIEF
CANNED FOODS
AND RAW OYSTERS
BALTIMORE, MD.
PLEASE RECEIPT AND RETURN TO
DAGUS MINES, PA.

Dicks-Rumberger Co., DuBois, PA**
Newcomb & Hand, Dover, DE
E.W. Milligan & Sons, Crisfield, MD
J. Langrall & Bro., Baltimore, MD
Billheads

GOODS BEING PERISHABLE ARE SHIPPED AT PURCHASER'S RISK.
Baltimore, Md., 190
Bought of
F. P. Coleman & Son
OYSTER PACKERS
118 CONCORD STREET.
PLEASE RECEIPT

J. D. GROVES & CO.
OYSTER & FRUIT PACKERS
PRIDE OF THE CHESAPEAKE
WHOLESALE SHIPPERS FRESH FISH, FRUITS AND PRODUCE
PRODUCE AND FRUITS OUR SPECIALTY.
P. O. BOX 904. Baltimore, 190
Sold to

CHALLENGE BRAND
Chas. Neubert & Co.
Oyster Packers
BALTIMORE OYSTERS
Baltimore, Md.

F. P. Coleman & Son, Baltimore, MD**
Full Moon Brand, Billhead;
J. D. Groves & Co., Baltimore, MD
Pride of the Chesapeake Brand, Billhead
Chas. Neubert & Co., Baltimore, MD
Challenge Brand, Letter

PERISHABLE GOODS SHIPPED AT PURCHASER'S RISK.
SWEET POTATOES
FLORIDA ORANGES
Baltimore, OCT 25 1892
Bought of JOHN W. NICOL,
OYSTER PACKER,
WHOLESALE
FRUITS AND PRODUCE,
CONTINENTAL BRAND
FRESH OYSTERS

Baltimore,
Bought of W. H. Lamon & Co.
Foreign FRUIT Domestic
Oyster and Fish Packers, Produce, &c.

ELK BRAND
F. BONHAGE & CO.
OYSTER PACKERS
AND SHIPPERS.
RAW OYSTERS
Fish and Produce.
BALTIMORE, MD.
SOLD TO

John W. Nichol, Baltimore, MD**
Continental Brand, Billhead;
W.H. Lamon & Co., Baltimore, MD Billhead;
F. Bonhage & Co., Baltimore, MD
Elk Brand, Billhead

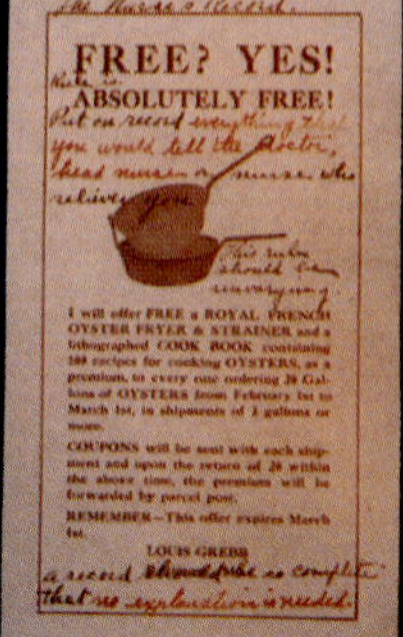

FREE? YES!
ABSOLUTELY FREE!

I will offer FREE a ROYAL FRENCH OYSTER FRYER & STRAINER and a lithographed COOK BOOK containing 100 recipes for cooking OYSTERS, as a premium, to every one ordering 20 Gallons of OYSTERS from February 1st to March 1st, in shipments of 3 gallons or more.

COUPONS will be sent with each shipment and upon the return of 20 within the above time, the premium will be forwarded by parcel post.

REMEMBER—This offer expires March 1st.

LOUIS GREBB

D.D. Mallory & Co., Baltimore, MD**
Diamond Brand Post Card;
Louis Grebb, Baltimore, MD
Free Oyster Fryer and Strainer Offer

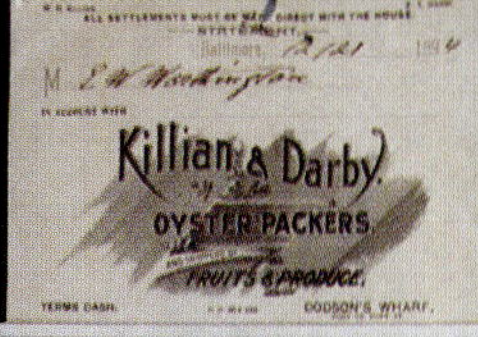
Killian & Darby
OYSTER PACKERS
FRUITS & PRODUCE.
TERMS CASH.
DODSON'S WHARF.

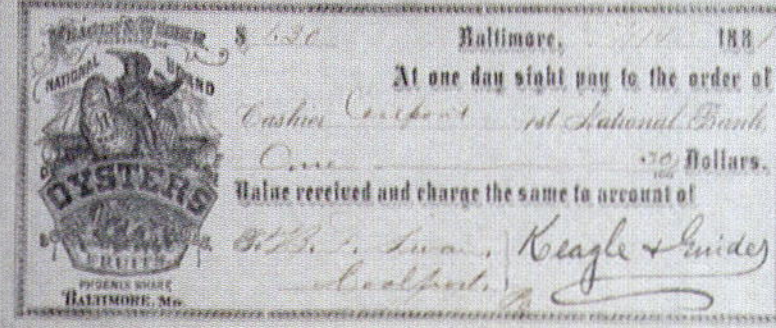
Baltimore, 188
At one day sight pay to the order of
Cashier 1st National Bank
Dollars.
Value received and charge the same to account of
Keagle & Guider

Baltimore,
BOUGHT OF F. BORDER'S SON & CO.
Oyster & Fruit Packers,
FULL MOON BRAND FRESH OYSTERS
FRESH FISH, FRUIT, VEGETABLES &c.
TERMS
BALTIMORE, MD. July 20, 19
Beadle & Co.,
Brockwayville, Pa.
clams package
PLEASE RECEIPT AND RETURN TO
OYSTERS FISH

Killian & Darby, Baltimore, MD;**
Keagle & Guider, Baltimore, MD
National Brand;
F. Border's Son & Co., Baltimore, MD
Full Moon Brand;
Planter's Trading Co., Baltimore, MD
Stag Brand;
Letterheads

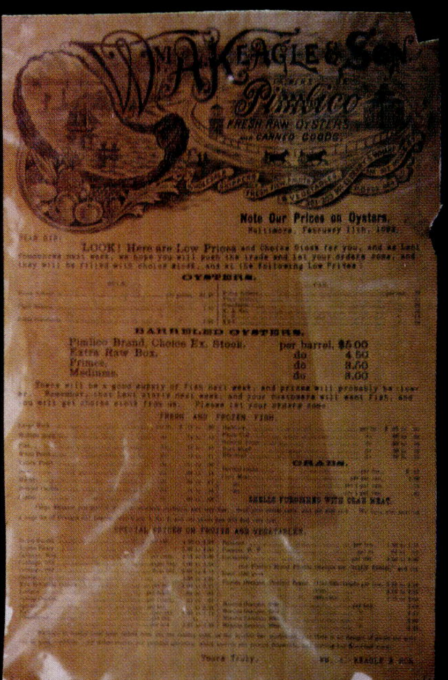
Wm. A. KEAGLE & SON
Pimlico
FRESH RAW OYSTERS AND CANNED GOODS.
Note Our Prices on Oysters.
OYSTERS.
BARRELED OYSTERS.
Pimlico Brand, Choice Ex. Stock. per barrel, $5.00
Extra Raw Box. do 4.50
Primes. do 3.50
Mediums. do 3.00
FRESH AND FROZEN FISH.
CRABS.
SHELLS FURNISHED WITH CRAB MEAT.
Yours Truly,
WM. A. KEAGLE & SON.

Wm. A. Keagle & Son, Baltimore, MD***
Pimlico Brand Price List

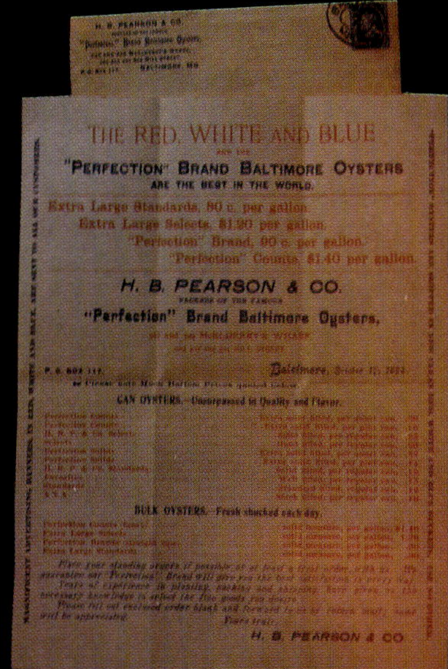
H. B. PEARSON & CO.
BALTIMORE, MD.
THE RED, WHITE AND BLUE
"PERFECTION" BRAND BALTIMORE OYSTERS
ARE THE BEST IN THE WORLD.
Extra Large Standards, 90 c. per gallon.
Extra Large Selects, $1.20 per gallon.
"Perfection" Brand, 90 c. per gallon.
"Perfection" Counts, $1.40 per gallon.
H. B. PEARSON & CO.
"Perfection" Brand Baltimore Oysters,
Baltimore,
CAN OYSTERS. Unsurpassed in Quality and Flavor.
BULK OYSTERS. Fresh shucked each day.
Yours truly,
H. B. PEARSON & CO.

H.B. Pearson & Co., Baltimore, MD***
1895 Perfection Brand Price List

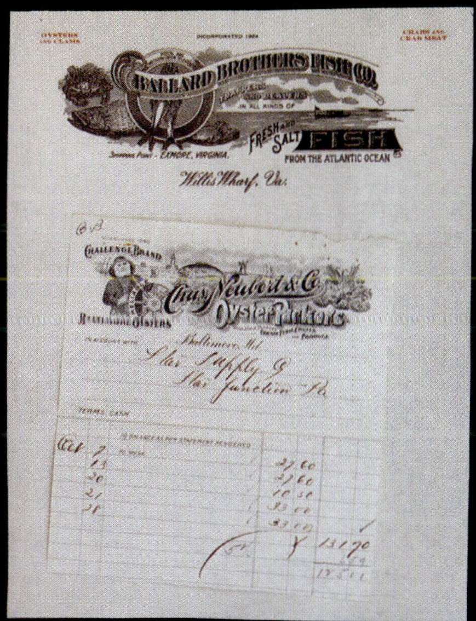
OYSTERS AND CLAMS
INCORPORATED 1924
CRABS AND CRAB MEAT
BALLARD BROTHERS FISH CO.
FRESH AND SALT FISH
FROM THE ATLANTIC OCEAN
Willis Wharf, Va.
CHALLENGE BRAND
Chas. Neubert & Co.
Oyster Packers
Baltimore, Md.
Star Supply Co
Star Junction Pa
TERMS: CASH

Ballard Brothers Fish Co., Willis Wharf, VA;**
Chas. Neubert & Co., Baltimore, MD
Challenge Brand;
Letterheads

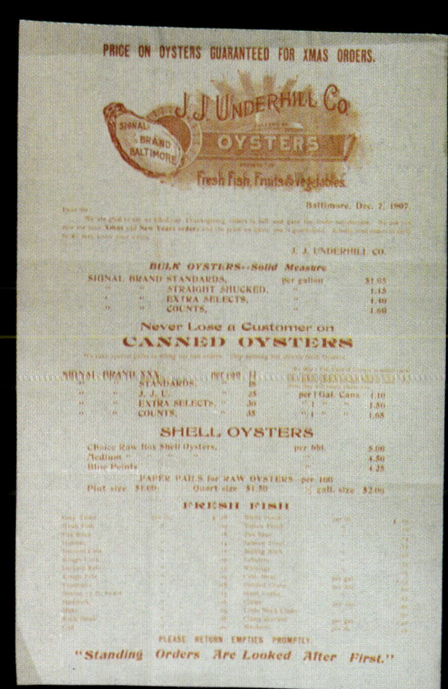
PRICE ON OYSTERS GUARANTEED FOR XMAS ORDERS.
J. J. UNDERHILL Co.
SIGNAL BRAND BALTIMORE
OYSTERS
Fresh Fish, Fruits & Vegetables.
Baltimore, Dec. 7, 1907
J. J. UNDERHILL CO.
BULK OYSTERS--Solid Measure
SIGNAL BRAND STANDARDS, per gallon
STRAIGHT SHUCKED,
EXTRA SELECTS,
COUNTS,
Never Lose a Customer on
CANNED OYSTERS
SHELL OYSTERS
PAPER PAILS for RAW OYSTERS
FRESH FISH
PLEASE RETURN EMPTIES PROMPTLY.
"Standing Orders Are Looked After First."

J.J. Underhill Co., Baltimore, MD***
Signal Brand Price List
Randy Shreck Collection

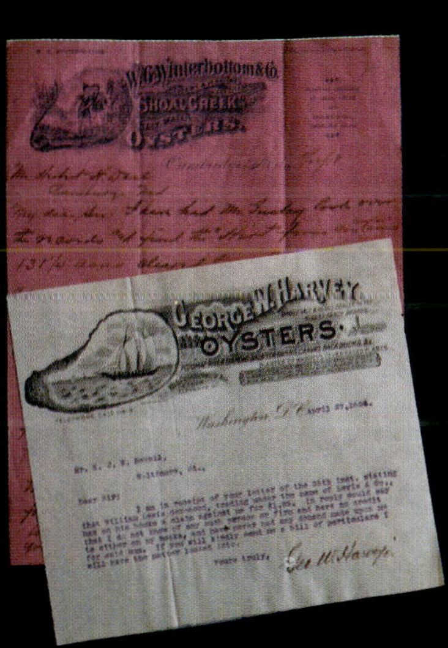
W. G. Winterbottom & Co.
SHOAL CREEK
OYSTERS
GEORGE W. HARVEY,
OYSTERS.
Washington, D. C.
Yours truly,
Geo W. Harvey

W. G. Winterbottom Co., Cambridge, MD*
Shoal Creek Brand;
George W. Harvey, Washington, DC
Letterheads

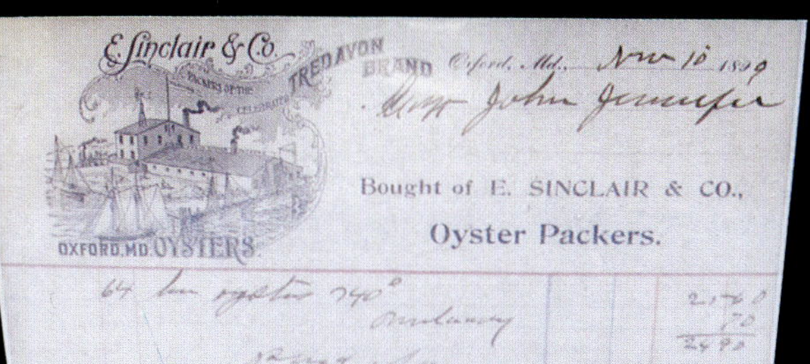
E. Sinclair & Co.
TRED AVON BRAND
Oxford, Md. Nov 10 1899
John Jennifer
Bought of E. SINCLAIR & CO.,
Oyster Packers.
OXFORD, MD. OYSTERS

E. Sinclair & Co., Oxford, MD***
Tred Avon Brand
Billhead
Mike and Eva Pinder Collection

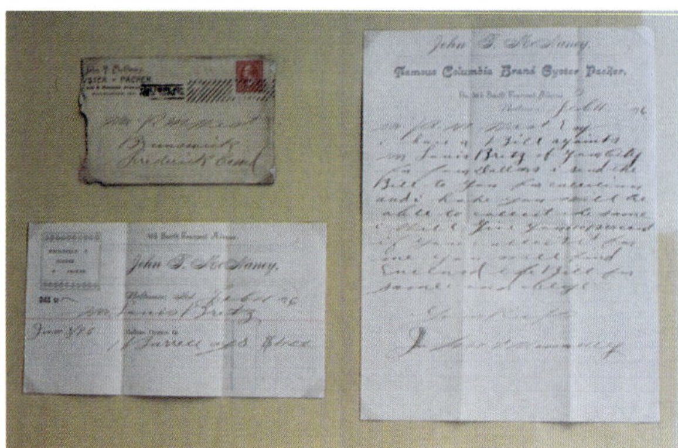

John W. McNaney, Baltimore, MD**
Columbia Brand, Framed Envelope
Bill and Letter
Roberta and Rudy Schmehl Collection

A.H.G. Mears, Wachapreague, VA**
Superlative Brand Oysters;
Donoho & Co., Seaford, DE;
Wm. L. Ellis & Co., Baltimore, MD
Star Brand;
Billheads

Union Oyster Co., Boston, MA**
Trade Card

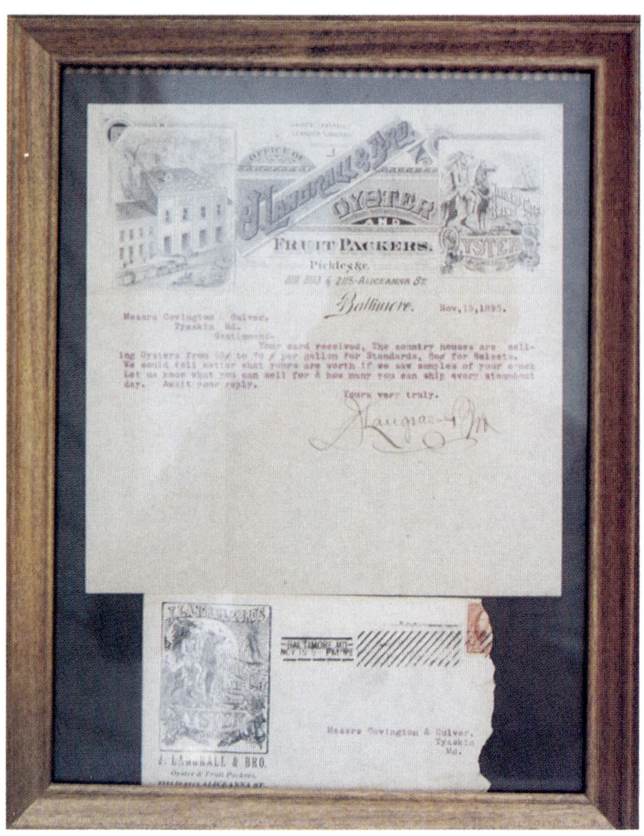

J. Langrall & Bro., Baltimore, MD**
Maryland Chief Brand, Framed Envelope
and Letter
Bill and Steve Dorrell Collection

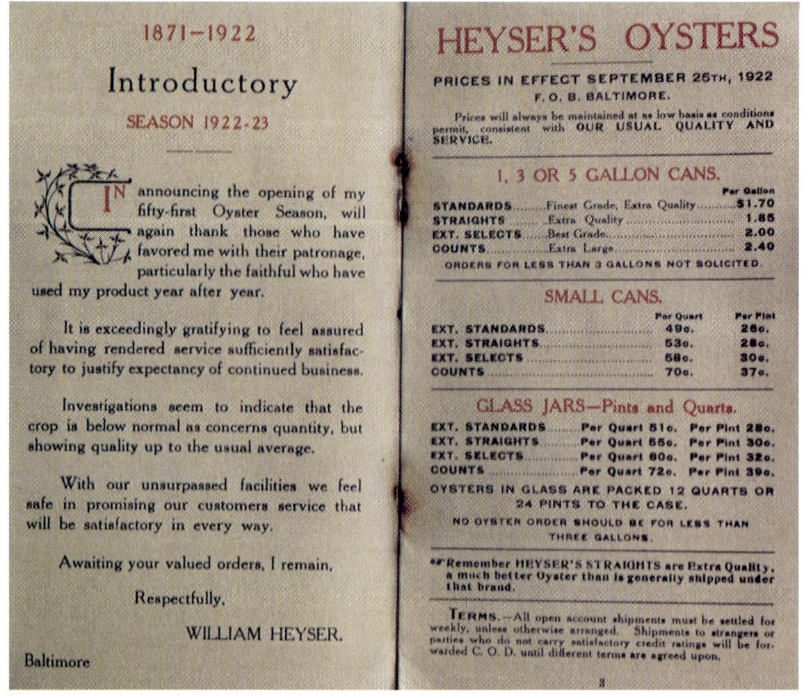

1871–1922

Introductory

SEASON 1922-23

In announcing the opening of my fifty-first Oyster Season, will again thank those who have favored me with their patronage, particularly the faithful who have used my product year after year.

It is exceedingly gratifying to feel assured of having rendered service sufficiently satisfactory to justify expectancy of continued business.

Investigations seem to indicate that the crop is below normal as concerns quantity, but showing quality up to the usual average.

With our unsurpassed facilities we feel safe in promising our customers service that will be satisfactory in every way.

Awaiting your valued orders, I remain,

Respectfully,

WILLIAM HEYSER.

Baltimore

HEYSER'S OYSTERS

PRICES IN EFFECT SEPTEMBER 25TH, 1922
F. O. B. BALTIMORE.

Prices will always be maintained at as low basis as conditions permit, consistent with OUR USUAL QUALITY AND SERVICE.

1, 3 OR 5 GALLON CANS.

		Per Gallon
STANDARDS	Finest Grade, Extra Quality	$1.70
STRAIGHTS	Extra Quality	1.85
EXT. SELECTS	Best Grade	2.00
COUNTS	Extra Large	2.40

ORDERS FOR LESS THAN 3 GALLONS NOT SOLICITED.

SMALL CANS.

	Per Quart	Per Pint
EXT. STANDARDS	49c.	26c.
EXT. STRAIGHTS	53c.	28c.
EXT. SELECTS	58c.	30c.
COUNTS	70c.	37c.

GLASS JARS—Pints and Quarts.

EXT. STANDARDS	Per Quart 51c.	Per Pint 28c.
EXT. STRAIGHTS	Per Quart 55c.	Per Pint 30c.
EXT. SELECTS	Per Quart 60c.	Per Pint 32c.
COUNTS	Per Quart 72c.	Per Pint 39c.

OYSTERS IN GLASS ARE PACKED 12 QUARTS OR 24 PINTS TO THE CASE.

NO OYSTER ORDER SHOULD BE FOR LESS THAN THREE GALLONS.

☞Remember HEYSER'S STRAIGHTS are Extra Quality, a much better Oyster than is generally shipped under that brand.

TERMS.—All open account shipments must be settled for weekly, unless otherwise arranged. Shipments to strangers or parties who do not carry satisfactory credit ratings will be forwarded C. O. D. until different terms are agreed upon.

3

Wm. Heyser Co., Baltimore, MD***
1922-23 Price List
Ronald L. Newcomb Collection

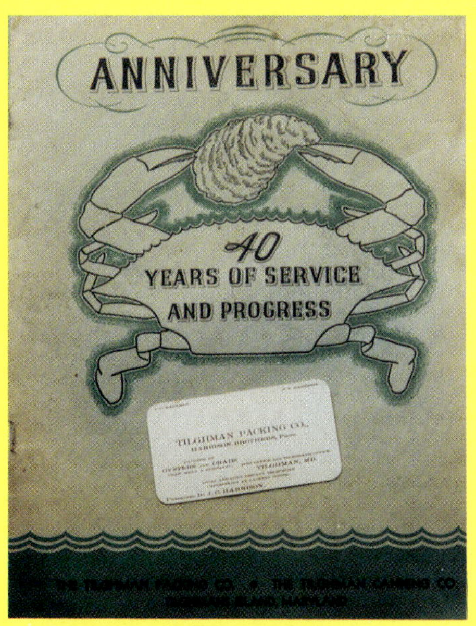

Tilghman Packing Co., Tilghman, MD***
Cover of Fortieth Anniversary Booklet
Courtesy of Nothing New Shop

Tilghman Packing Co., Tilghman, MD
Display of Products from Anniversary Booklet
Note Pearl Necklaces Used as Gifts
Courtesy of Nothing New Shop

The Daily Pioneer Press.

ESTABLISHED IN 1874.

R. F. JONES,

Packer of the Celebrated and Popular

GOLD SEAL BRAND

FRESH OYSTERS.

Superior quality Fresh Oysters, standard filled cans, eight solid pints to the gallon.

R. F. JONES'

GOLD SEAL SELECT,

GOLD SEAL STANDARD,

R. F. J. MEDIUM.

AM NOW SHIPPING IN CAR LOADS ONLY

Fresh Salt Water Fish of all kinds at Very Low Winter Rates.

R. F. JONES,

BALTIMORE, MINNEAPOLIS AND WINNIPEG.

R.F. Jones, Baltimore, MD**
Newspaper Ad Daily Pioneer Press,
Minneapolis, MN December 1881

Tilghman Packing Co., Tilghman, MD
Display of Products from Anniversary Booklet
Courtesy of Nothing New Shop

The Great American Tea Co., New York, NY**
Trade Card, Oyster's Two: "How Dreadfully Hot It Is."
"Yes, I Feel Quite In A Stew."

The Great American Tea Co., New York**
Trade Card, "Oh! Bivalve I Fear These Cannibals Are After Us."
"Do Not Fear Shelly Dear There is No R in This Month."

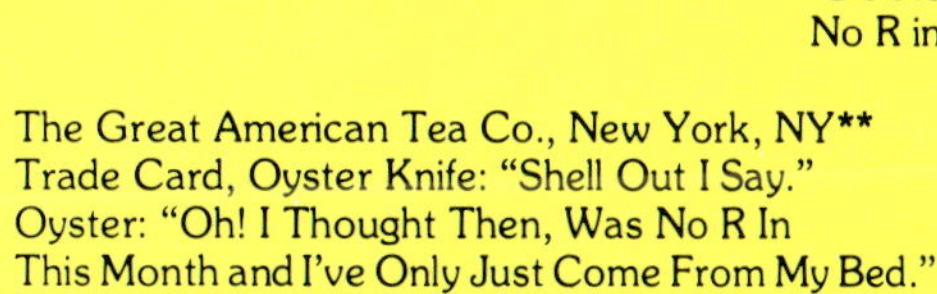
The Great American Tea Co., New York, NY**
Trade Card, Oyster Knife: "Shell Out I Say."
Oyster: "Oh! I Thought Then, Was No R In This Month and I've Only Just Come From My Bed."

Taunton Iron Works, Taunton MA**
Trade Card, Mr. Lemon: "Can I Do Anything For You Miss Shell?"
Miss Shell: "No Thank You I Can Get Along Without Lemon Aide."

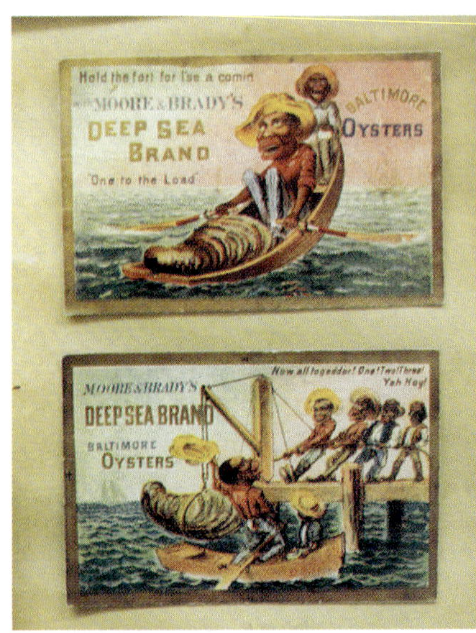

Moore & Brady, Baltimore, MD**
Deep Sea Brand, Two Trade Cards
Carlton and Mary Riggin Collection

W.R. Barnes & Co., Baltimore, MD**
Big Gun Brand;
D. D. Mallory & Co., Baltimore, MD
Diamond Brand;
A. Booth, Baltimore, MD
Oval Brand, Trade Cards
Joe Sechrist Collection

James E. Stansbury Co., Baltimore, MD**
Pioneer Brand, Trade Card
Butch and Jackie Cheezum Collection

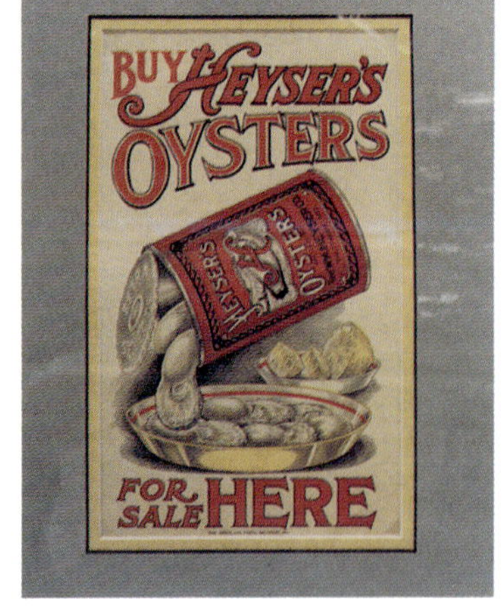

Wm. Heyser & Co., Baltimore,
MD****
Advertising Sign
Courtesy of Black Swan Antiques

D.E. Foote, Baltimore, MD**
Compass Brand, Trade Card
Randy Shreck Collection

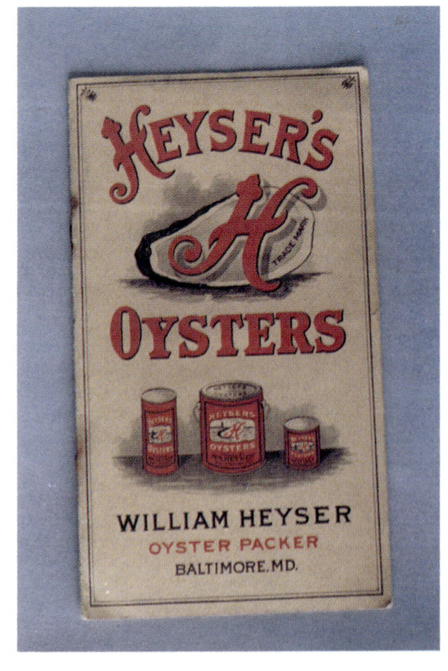

William Heyser, Baltimore, MD****
Advertising Sign
Ronald L. Newcomb Collection

Oyster Cracker Advertisement***
Carlton and Mary Riggin Collection

Mace, Woolford & Co., Cambridge, MD***
Advertising Post Card 1908
Ronald L. Newcomb Collection

Crisfield, MD Post Cards**
Courtesy of Country Quest Antiques

Jas. Hubbard & Son, Baltimore, MD****
Plymouth Rock Advertising Sign
Ronald L. Newcomb Collection

Louis Grebb, Baltimore, MD****
Belle Brand Calendar
Ronald L. Newcomb Collection

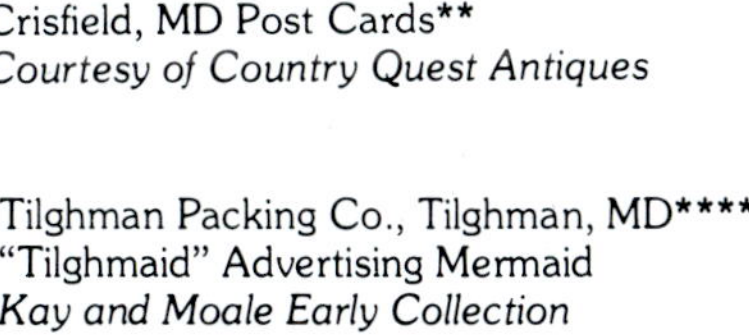
Tilghman Packing Co., Tilghman, MD****
"Tilghmaid" Advertising Mermaid
Kay and Moale Early Collection

Tonging License Receipt 1898**

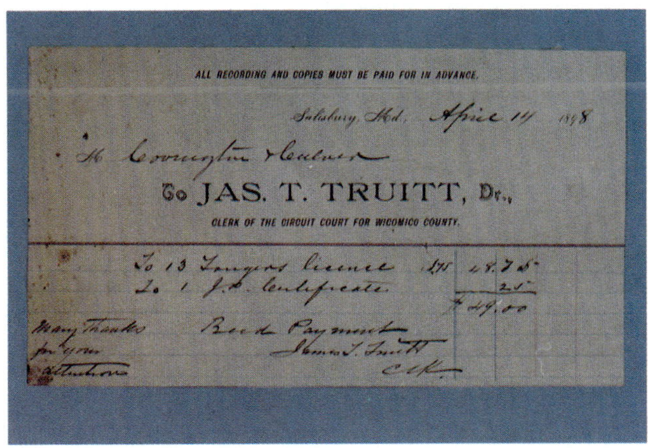
ALL RECORDING AND COPIES MUST BE PAID FOR IN ADVANCE.

Salisbury, Md. April 14 1898

To JAS. T. TRUITT, Dr.,

CLERK OF THE CIRCUIT COURT FOR WICOMICO COUNTY.

To 13 Tongers licence 48.75
To 1 J.P. Certificate .25
49.00

Many Thanks for your Attention
Recd Payment
James T. Truitt
Clk.

John T. McNaney, Port Norris, NJ***
Advertising Thermometer
Joe Sechrist Collection

Woodfield Fish & Oyster Co., Galesville, MD***
Set of Playing Cards
Mike and Eva Pinder Collection

Set of Ceramic Oyster Baking Dishes**
Courtesy of Harris Crab House

Shelter Island Oyster Co., Greenport, NY**
Set of Oyster Shell Baking Dishes
Stewart Ewell Collection

Independent Can Co., Baltimore,

50th Anniversary Cooler
Courtesy of Harris Crab House

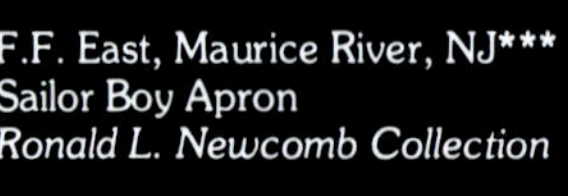

F.F. East, Maurice River, NJ***
Sailor Boy Apron
Ronald L. Newcomb Collection

F.F. East, Maurice River, NJ**
Coin Purse
Ronald L. Newcomb Collection

The Lowndes Oyster Co., South Norwalk, CT***
Cigarette Lighter
Ronald L. Newcomb Collection

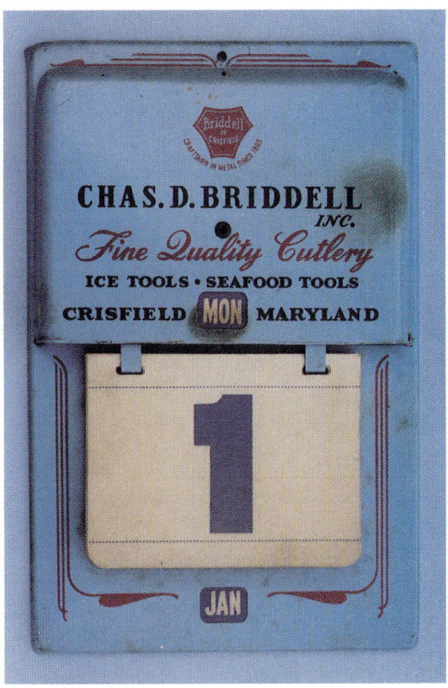

Chas. D. Briddell, Crisfield, MD**
Perpetual Calendar

Sealshipt Oyster System, Inc.***
South Norwalk, CT, Watch Fob
Ronald L. Newcomb Collection

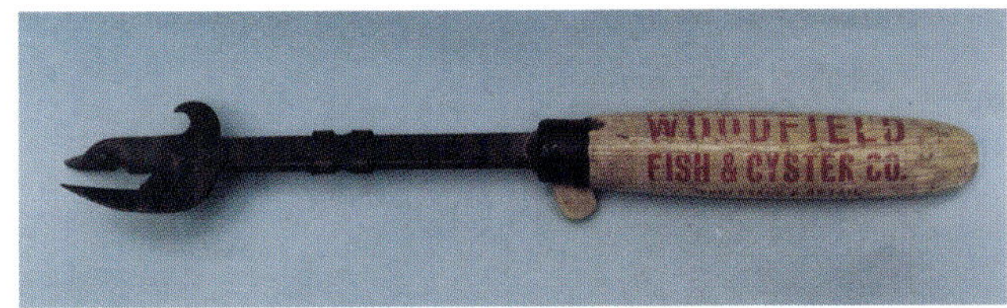

Woodfield Fish & Oyster Co., Galesville, MD**
Can Opener
Gary and Sharon Campbell Collection

Chas. Neubert & Co., Baltimore, MD**
Can and Bottle Opener
Gary and Sharon Campbell Collection

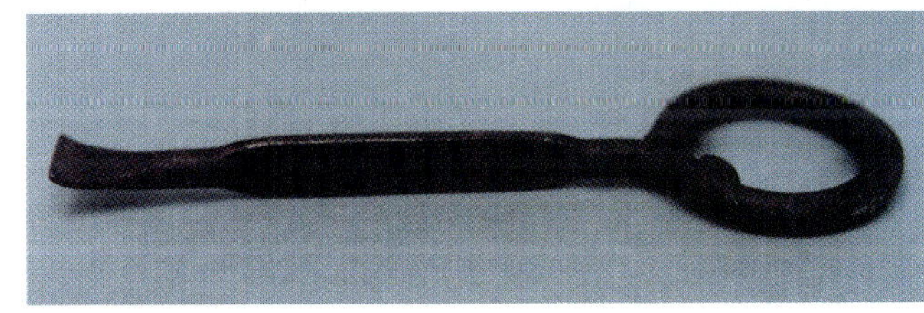

R. W. Strickler, York PA**
Can Opener
Drew W. Landis Collection

Chas. Neubert & Co., Baltimore, MD**
Ruler
Ronald L. Newcomb Collection

Stowman Bros., Maurice River, NJ**
Captain Jacks, Advertising Pencil

The Oyster Shell Products Co., Philadelphia, PA**
Ruler

Frank Hittle & Co. Baltimore, MD***
Paperweight
Ronald L. Newcomb Collection

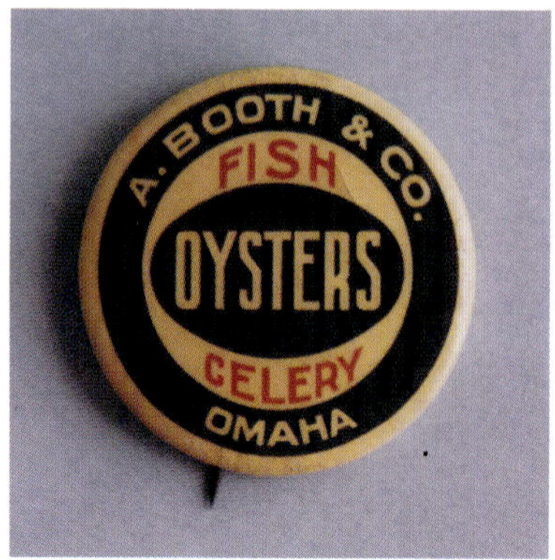

A. Booth & Co., Omaha, NEB**
Advertising Pinback
Ronald L. Newcomb Collection

Advertising Pinback**
Ronald L. Newcomb Collection

Wilkinson Bros.**
Advertising Pinback
Ronald L. Newcomb Collection

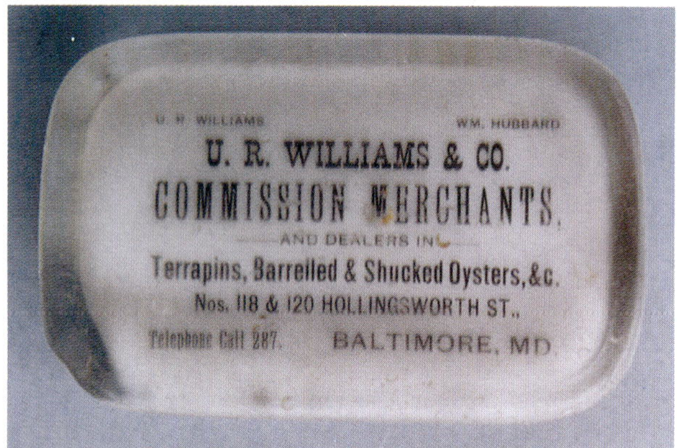

U.R. Williams & Co., Baltimore, MD***
Paperweight
Ronald L. Newcomb Collection

Keeling-Easter Co., Norfolk, VA***
Paperweight
Ronald L. Newcomb Collection

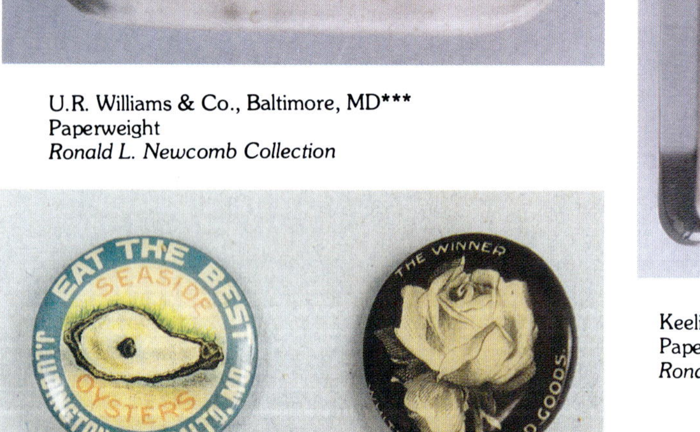

J. Ludington & Co., Baltimore**
Seaside Oyster Pinback
Joe Sechrist Collection

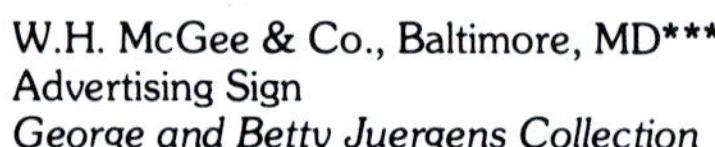

W.H. McGee & Co., Baltimore, MD***
Advertising Sign
George and Betty Juergens Collection

Lord Mott Co. & R.E. Roberts Co. Sign,****
Note They Shared Offices
Carlton and Mary Riggin Collection

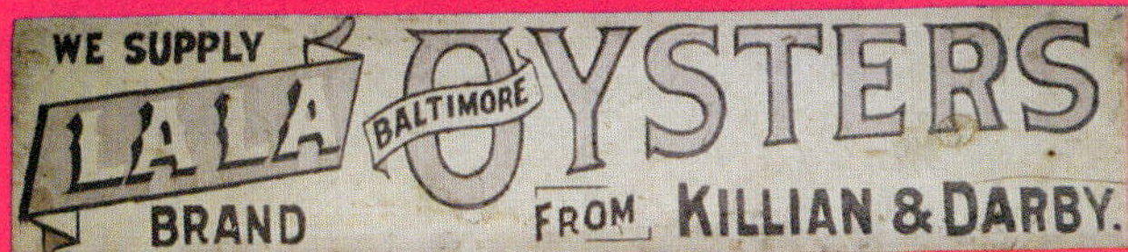

Killian & Darby, Baltimore, MD***
La La Brand, Advertising Sign
Ronald L. Newcomb Collection

Peterson Packing Co., Port Norris, NJ**
Advertising Sign
Ronald L. Newcomb Collection

W.H. McGee Co., Baltimore, MD****
Advertising Sign
Ronald L. Newcomb Collection

H.C. Rowe & Co.***
Advertising Sign
Ronald L. Newcomb Collection

A. Phillips & Co., Cambridge, MD****
Advertising Sign
Ronald and Peggy Rue Collection

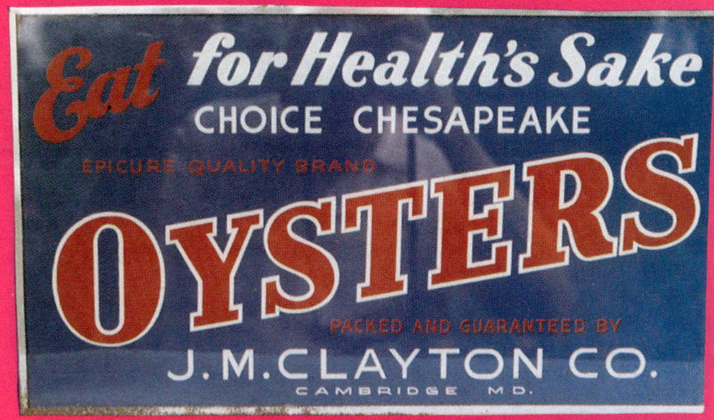

J.M. Clayton Co., Cambridge, MD***
Advertising Sign
Ronald L. Newcomb Collection

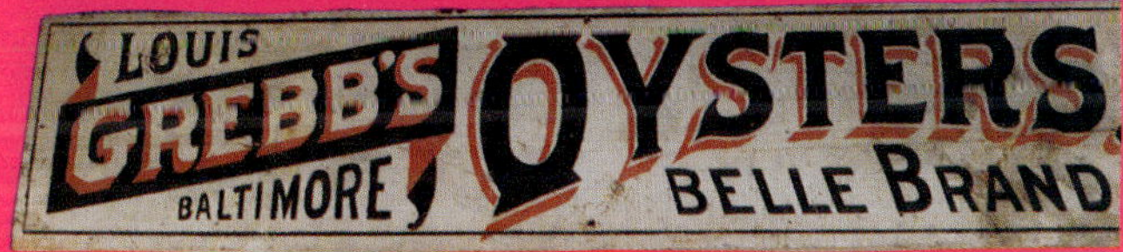

Louis Grebb, Baltimore, MD***
Advertising Sign
Ronald L. Newcomb Collection

Cartwright's Robbins Island Oysters***
Advertising Sign
Ronald L. Newcomb Collection

American Wine Co., St. Louis, MO***
Cook's Wine Advertising Panel
Ronald L. Newcomb Collection

American Wine Co., St. Louis, MO****
Cook's Wine Advertising Tray
Carlton and Mary Riggin Collection

Platt & Co., Baltimore, MD****
Tiger Brand Sign
Ronald L. Newcomb Collection

The Leib Packing Co., Baltimore, MD***
Sun Brand Advertising Sign
Ronald L. Newcomb Collection

Bluepoints Co. Inc. West Sayville, NY***
Sealshipt Advertising Sign
Ronald L. Newcomb Collection

C.H. Lighthiser, Baltimore, MD****
Elephant Brand Advertising Sign
George and Betty Juergens Collection

Shreve's Bros., Chincoteague, VA****
Original Artwork for Can

Pabst Brewing Co.,***
Advertising
Courtesy of Black Swan Antiques

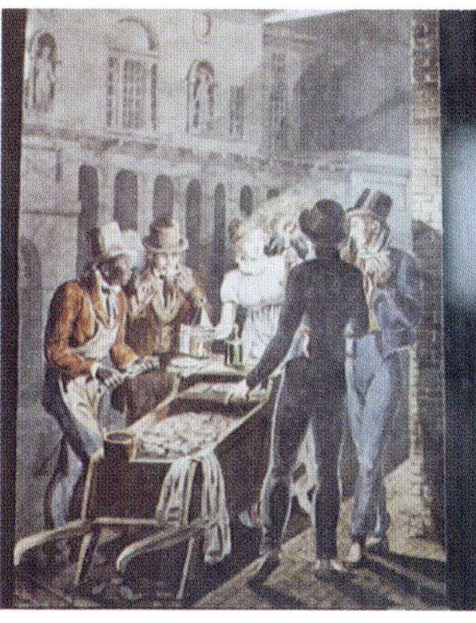

Picture of Oyster Vendors***
Carlton and Mary Riggin Collection

The Leib Packing Co., Baltimore, MD***
Framed Display of Oyster Shell and Sign
Ronald L. Newcomb Collection

Moyer & Son, Souderton, PA***
Oyster Shell Bag Stencil
Rudy and Roberta Schmehl Collection

MUSEUMS

There are a number of museums that focus on the oyster. They have interpretative exhibits showing the oyster industry as it applied to the shucking and marketing of oysters. The Chincoteague Oyster and Maritime Museum shows how it was done on the seaside at Chincoteague, Virginia. The Maritime Museum in Solomons, Maryland presents the J. C. Lore oyster house as a working oyster shucking and marketing enterprise and the Baltimore Museum of Industry near the Inner Harbor, has a working shucking house set up to provide school children with a chance to find out what a job in the oyster industry was like, including pay with tokens, redeemable in the company store.

Calvert Marine Museum
J.C. Lore Oyster House
Solomons, MD

Oyster and Maritime Museum
Chincoteague, MD

Bibliography

Bolitho, Hector, Editor, **The Glorious Oyster**, New York, Horizon Press, 1961

Brooks, William K., **The Oyster,** Johns Hopkins Press, 1891

Burton, R. Lee, Jr., **Canneries of the Eastern Shore**, Tidewater Publishers, Centerville, MD, 1986

Chowning, Larry, **Harvesting the Chesapeake**, Tidewater Publishing, Centerville, MD, 1990

Hedeen Robert, **The Oyster**, Tidewater Publishers, Centerville, MD, 1986

Johnson, Paula, Editor, **Working the Water**, University of Virginia Press, Charlottesville, VA, 1988

Kochiss, John, **Oystering from New York to Boston**, Wesleyan Univ. Press, 1974

May, Earl Chapin, **The Canning Clan**, Macmillan Co., 1937

Parks, Fred, **Oysterhouse Cookbook**, Allentown PA, 1985

Robinson, Robert, **The Illustrious Oyster Illustrated**, Sussex Prints, Georgetown, DE, 1983

Schenkman, David E. **Maryland Merchant Tokens**, Maryland Token and Medal Society, Inc., Baltimore, MD, 1986

Picture of George Christy Oyster Packing
Ralph and Betty Tull Collection

Brands List

Brands of oyster are placed in boldfaced type, followed by the packing company name and the town of operation.

Abbott
Abbott Bros.
Weems, VA

Acme
Applegarth, C.L. & Sons
Baltimore, MD

Acme
Hutchinson, J.G. & Co.
Ottum, WA

Adam's
Adam's, W.J.
Chincoteague, VA

Allen's
Allen's Oyster House
Coles Point, VA

Alligator
Ruge Bros. Canning Co.
Apalachicola, FL

American Beauty
Anticich Canning Co.
Biloxi, MS

AMW
Woodburn, H.M. & Son
Solomons, MD

Apalachicola Bay
Kirvin Bros. Sea Food Co.
Apalachicola, FL

Apco
Adams Packing Corp.
Grimstead, VA

Arrow
Lansburgh, J.J. & Co.
Baltimore, MD

Atkin's
Grahams Seafood Co.
Coden, AL

B & J
Madison Seafood Co.
Madison, MD

B & L
Bivalve Oyster Packing Co.
Bivalve, MD

B & M
Bunting's, B.M. Oyster House
Gloucester, VA

B & S
B & S Fisheries, Inc.
Fall River, MA

Back River Seafood
Cook's Seafood
Hampton, VA

Banner
National Packing Co., The
White Stone, MD

Banner
National Packing Co.
Baltimore, MD

Bay
York River Seafood Co.
Seaford, VA

Bay Bridge
Ruth, H.W. & Son
Grasonville, MD

Bay City
Bay City Foods
Baltimore, MD

Bay Shore
Harris, W.H. Seafood
Chester, MD

Bayou Rose
Morgan City Packing Co.
Houma, LA

Belle
Grebb, Louis
Baltimore, MD

Belleview
Valliant, W.H. & Bro.
Belleview, MD

Bennett's
Bennett's Seafood Co.
Pass Christian, MS

Bevins
Bevin's Oyster Co.
Kinsale, VA

Bewdly
Pittman, W.R. & Sons
Somers, VA

Big "C"
Coulbourn, J.C. Co.
Baltimore, MD

Big Gun
Barnes, W.R. & Co.
Baltimore, MD

Big H
Hale Seafoods
Morattico, VA

Big L Fish Hawk
Lumpkin, R.H. & Son
Kilmarnock, VA

Big S
Slaughter, C.T.
Morattico, VA

Billup's
Billups, H.K.
Matthews, VA

Billup's
Billup's, H.K.
Motorun, VA

Biloxi
Biloxi Canning & Packing Co.
Biloxi, MS

Black Pearl
Crockett & Forbush
Crisfield, MD

Black Swan
Leatherbury Bros.
Shadyside, MD

Bleakhorn
Nansemond Adams Oyster Co.
Crittenden, VA

Blue Cold
Quebec United Fishermen
Montreal, PQ

Blue Cross
Travers Bros. Co.
Baltimore, MD

Blue Seal
Bluepoints Co. Inc.
West Sayville, NY

Blue Seal
Ipswich Shellfish Co.
Ipswich, MA

Blue Star
Carnabuci Seafoods, Inc.
Grasonville, MD

Bluepoints' Bay Oysters
Bluepoints Co., Inc.
Cambridge, MD

Bluff Point
Rappahannock Oyster Co.
Kilmarnock, VA

Bluffton
Graves, J.S. Jr.
Bluffton, SC

Booth
Booth Fisheries Co.
Chicago, IL

Booth
Booth Fisheries Co.
Crisfield, MD

Boss
Beckwith, H. & Co.
Baltimore, MD

Boyer's
Boyer, W.W. & Co.
Baltimore, MD

Brannocks
Brannock, H.I. & Co.
Cambridge, MD

Briny Deep
Boak, R.B. Co.
St Paul, MN

Bull Bay
Glines Seafood
Awendaw, SC

Bull's Head
Gibbs Preserving Co.
Baltimore, MD

"C"
Christy, George A. & Son
Crisfield, MD

Calvert
Denton & Rogers
Broomes Island, MD

Calvert's
Calvert's Shellfish Co.
Grasonville, MD

Camel
unknown
Baltimore, MD

Cannon's Quality
Cannon, I.F. & Son
Cambridge, MD

Cap'n Gollott's
Gollott, E.R.
Biloxi, MS

Cap'n Hank's
Hank's Seafood Co.
Easton, MD

Capt. Sol's
Independent Fish Co.
Birmingham, AL

Cape Chester
C & C Shellfish Co.
Rock Hall, MD

Cape May
Port Norris Oyster Co.
Port Norris, NJ

Cape May
Stowman Bros.
Maurice River, NJ

Cape May Salts
East Point Oyster Co.
Cape May Court House, NJ

Capitol
Capitol Fish Co.
Atlanta, GA

Capt. Boyds
Saunders, C.P. & Son
Millenbeck, VA

Capt. Colliers
Randolph's Seafood
Coden, A.

Capt. Fisher
Fisher, Lance G. & Oyster Co.
Sanford, VA

Captain Pitre's Best
Golden Meadow Oyster House
Golden Meadow, LA

Castle
Castle Oyster Co.
Baltimore, MD

Caviar
Unknown

Cedar Creek
Cedar Creek Packing Co.
Wenona, MD

Cedar Points
Tallmadge Bros. Inc.
South Norwalk, CT

Challange
Lanfair, H.S. & Co.
Baltimore, MD

Challange
Neubert, Chas. W. & Co.
Baltimore, MD

Chase
Chase, W.L. Co.
Norfolk, VA

Chef
Jacobs, Wm. & Sons
Baltimore, MD

Cherrystone
Steelman, E.J.
Cheriton, VA

Chesapeake Bay
Chesapeake Bay Oyster Exchange
Baltimore,. MD

Chesapeake Bay
Miles, J.H. & Co.
Norfolk, VA

Chief Engelhard
Engelhard Shrimp, Fish & Oyster Co.
Engelhard, NC

Chincoteague Bay
Collins, Nelson P.
Cincoteague, VA

Chincoteague Island
Stubbs, Reginald Seafood
Chincoteague, VA

Choice of Chesapeake
Harding Seafood Co.
Weems, VA

Choice of Chesapeake Bay
Ferguson, J.W. Seafood Co.
Remlick, VA

Christy's
Christy, George A. & Son
Crisfield, MD

Clifton's
Clifton, J.S.
Shallote, NC

Coast-Pact
Port Norris Oyster Co.
Port Norris, NJ

Columbia
McNaney Oyster Co.
Baltimore, MD

Columbian
Wrightson's
Baltimore, MD

Compass
Big Bear Stores
Columbus, OH

Continental
Foote, D.E. & Co.
Baltimore, MD

Cook's
Cook's Seafood Co.
Bena, VA

Corvette
Quebec United Fisherman
Montreal, Quebec

Cream O'Sea
Brown, R.F. Sea Food Co.
Lansing, MI

Crescent
Crescent Sea Food Co.
Baltimore, MD

Crisfield
Crisfield Packing Co.
Crisfield, MD

Crystal Bay
Shelmore Oyster Co.
Charlestown, SC

Curley's Pearl of Perfection
Curley Packing Co.
Colonial Beach, VA

D & D

Dean Seafood
Montross, VA

Darling's
Darling, J.S. & Son
Hampton, VA

Daufuski
Maggioni, L.P. & Co.
Savannah, GA

Davy Crockett
Crockett Seafood inc.
Irvington, VA

De Jean's
De Jean Packing Co.
Biloxi, MS

Deep River
Crisfield Sea Food Co.
Baltimore, MD

Deep Rock
Coulbourn, N.R.
Crisfield, MD

Deep Sea
Moore & Brady
Baltimore, MD

Deer Head
Dunbar-Dukate Co.
New Orleans, LA

Deer Island
Baltimore & Chesapeake Oyster Co.
Baltimore, MD

Del-Mar-Va
Ward Oyster Co.
Crisfield, MD

Delicious
Cornwell Seafood Co.
Irvington, VA

DeLuxe
White, J.H. Co.
Baltimore, MD

Diamond
Mallory, D.D. & Co.
Baltimore, MD

Diamond K
Kelley, R.L.
Atlantic, VA

Diamond Point
Still, Geo. M. & Co.
New York, NY

Diamond Shoals
Fulcher, Clayton Seafood Co.
Atlantic, NC

Dill's
Dill's Seafood
Bridgetown, NJ

Dog's Head
Martin Wagner Co.
Baltimore, MD

Double DD
David Davies
Columbus, OH

Double "S"
Southern Sea Food Co.
Baltimore, MD

Dryden's Quality
Dryden, Carol & Co.
Crisfield, MD

Duke of Gloucester
Gloucester Seafood Packing Co.
Bena, VA

Dutch
Griffin, B.A. Co.
Milwaukee, WI

Dutch Cove
Webb, E.I. & Co.
Weems, VA

Eagle
Eagle Packing Co.
Baltimore , MD

East Coast
Lynch Fish Co.
Cincinnati, OH

Eastpoint
Greenwich Oyster Co.
Greenwich, NJ

Ed. Martin's
Martin, Ed. Seafood Co.
Westwego, LA

Egg Island
Ballard Brothers Fish Co.
Willis Wharf, VA

Egg Island
Ballard Fish & Oyster Co.
Norfolk, VA

Elephant
Lighthiser, C.H.
Baltimore, MD

Elk
Bonhage, F. & Co.
Baltimore, MD

Elliott's
Elliott, G.T., Inc.
Hampton, VA

Epicure
Clayton, J.M. Co.
Cambridge, MD

Evans
Evans, V.L. & Co.
Crisfield, MD

Everbest
Stowman Bros.
Maurice Rivers, NJ

Express
McWilliams, H. & Co.
Baltimore, MD

"F"
Flint, M.G. Mercantile Co.
Denver, CO

Famous
United Shellfish Co.
Ipswich, MA

Famous Bivalve Seafood
Bivalve Seafood Co.
Bivalve, NJ

Famous Pautunxent River
Lore, J.C. & Sons
Solomons, MD

Florida Bahama
Florida Bahama Seafood Co.
Boynton Beach, FL

Foote's Best
Foote, D.E. & Co.
Baltimore, MD

Ford's Gulfstream
Ford's Sea Products Corp.
Mobile, AL

Forest
Forest Oyster Co.
Atlantic, VA

Forty [40] Fathom
General Seafood, Ltd.
Halifax, NS

Fres-Shore
Kroger Stores
Cincinnati, OH

Frontier
Nove-McCord Mecantile Co.
St. Joseph, MO

Full Moon
Coleman, F.P. & Son
Baltimore, MD

Full Moon
Watson, G. & Co.
Savannah, GA

Full Packt
Holmes, Ernest C.
Waterloo, NY

Full Packt
Neubert, Charles
Baltimore, MD

G & E
Mt Vernon Packing Co.
My Vernon, MD

Garden State
Garden State Oyster Co.
Port Norris, NJ

Gem of the Sea
Apple Cupboard Grocery Co.
Memphis, TN

Gold Seal
Jones, R.F.
Baltimore, MD

Gollott's
Gollott & Son Seafood Co.
Biolxi, MS

Goose Creek
Mace, Woolford & Co.
Cambridge, MD

Grasso
Grasso, Joe & Son
Galveston, TX

Grassy Hammock
Lovejoy, Frederick F.
East Norwalk, CT

Green Vale
Conrad, E.J. & Son Seafood
Lancaster, VA

Green Vale
Dobyns, R.E.
Monaskon, VA

Green's
Green's Oyster Co.
Shallotte, NC

Guiding Light
Ashburn, Oscar & Son
Weems, VA

Gulf's Best
Gulf's Best Food
Bayou La Batre, AL

H & B
Howeth, C.W. & Bro.
Crisfield, MD

Haines
Haines Oyster Co.
Seattle, WA

Hampton Roads
Deltaville Fish & Oyster Co.
Deltaville, VA

Handy's
Handy, John T. Co.
Crisfield, MD

Harbor Cove
Chesapeake Shellfish Co.
Sherwood, MD

Harrington
Harrington, Philip J. & Son
Secretary, MD

Harrison
Harrison Oyster Co.
Tilghman, MD

Harvard
Whitman, Ward & Lee Co.
Boston, MA

Health Seal
Smalley, Kivlan, Anthank
Boston, MA

Heart of America
Christopher Mercantile Co.
Kansas City, MO

Helsby
Helsby Seafood Co.
Crisfield, MD

Heyser's
Heyser, Wm. & Co.
Baltimore, MD

High Tide Brand
Stowman Bros.
Maurice River, NJ

Hilton Head
Hutchins, H.D.
Hilton Head, SC

Hogg's
Hogg, F.C.
Gloucester Point, VA

Honga
White & Nelson
Cambridge, MD

Howard Johnson's
Howard Johnson's
Wollaston, MA

Hughlett Point
Hughlett Seafood
Kilmarnock, VA

Humpty-Dumpty
Burgess-Humphrey's Canning Co.
New Orleans, LA

Indian Creek
Rappahannock Oyster Co.
Kilmarnock, VA

Irma
McCaddin, Wm. B. & Co.
Baltimore, MD

Irvington
Irvington Fish & Oyster Co.
Irvington, VA

Isaac Fass
Fass, Isaac, Inc.
Portsmouth, VA

Island Rock
Rhodes, Grady Seafood
Saxis, VA

Islander
Islander Seafoods, Inc.
Grasonville, MD

J.B. & Sons
Brinkley, J.B. & Sons
Baltimore, MD

Jeff Johnson's
Johnson, Jeff
Geron Bay, AL

Jefferies
Jefferies' Oyster Farms, Inc.
Bivalve, NJ

Jersey
DuBois, E.C. & Co.
Bivalve, NJ

Jersey Cape
Phillips Seafood Packing Co.
Port Norris, NJ

Jersey's Best
Robbins Brothers
Port Norris, NJ

Jones
Jones, Thos. E.
Cambridge, MD

Jumbo
Miller Bros. & Co.
Baltimore, MD

K & B
Thurston & Kingsbury
Bangor, ME

Kambur's Specials
Kambur, J. Co.
New Orleans, LA

Kellum
Kellum, Ellery
Weems, VA

Kent Island
Harris, A.C. Co.
Chester, MD

Keystone
Durm, Wm. B. & Co.
Baltimore, MD

Keystone
Stirling Oyster Co.
Detroit, MI

King Carter
Virginia Seafoods, Inc.
Irvington, VA

King Cole
King Cole Co.
Omaha, NB

King's Choice
King's Choice Foods
Los Angeles, CA

Kirkpatrick's
Kirkpatrick, Allen & Co.
Dover, DE

Kirkpatrick's
Kirkpatrick, Allen & Co.
Rehobeth, DE

La La
Killian & Darby
Baltimore, MD

La Mariniere
Quebec United Fishermen
Montreal, Quebec

Lady Adams
Mebus & Drecsher
Sacramento, CA

Lawson
Lawson Oyster Co.
Crisfield, MD

Leonard's
Leonard, I.L. & Co.
Cambridge, MD

Liberty
Ivens & Hudson
Rock Hall, MD

Lightning Express
Thurston, Russell & Co.
Annapolis, MD

Long Bar
Oxford Packing Co.
Oxford, MD

Louisiana Bayou
De Sanka Oyster House
Amite, LA

Lovely Lady
Leonard, John H.
Baltimore, MD

Lux
Lux, E.A. & Co.
Baltimore, MD

M & M
Martina & Martina
New Orleans, LA

M & V
Unknown
Crisfield, MD

Majestic
Atlantic Packaging Co.
Baltimore, MD

Manokin River
Bozman, Harold Seafood
Upper Fairmount, MD

Mar-Va
Dize, E.R. & Co.
Crisfield, MD

Markos
Rock Hall Clam & Oyster Co.
Rock Hall, MD

Mary K
Diggs, E.H. & Son
Grasonville, MD

Maryland
Myer, T.J. & Co.
Baltimore, MD

Maryland Beauty
Roberts, R.E. Co.
Baltimore, MD

Maryland Chief
Langrall, J. & Bro.
Baltimore, MD

Maryland House
Nanticoke Seafood Co.
Nanticoke, MD

Maryland's Finest
Todd Seafoods, Inc.
Cambridge, MD

Maryland's Pride
Rochester, Harry Co.
Baltimore, MD

McCready
McCready Bros.
Chincoteague, VA

McNaney
McNaney, John T.
Baltimore, MD

Meredith
Meredith & Meredith
Wingate, MD

Merit
Ragland, Potter Co.
Tennessee

Merit
Unknown

Mermaid
Neubert, Chas. & Co.
Baltimore, MD

Mermaid Oysters
Mebus & Drescher Co.
Sacramento, CA

Metompkin
Metompkin Bay Oyster Co.
Crisfield, MD

Mexican Gulf
Mexican Gulf Fisheries
Coden, AL

Miah-Maull
Peterson Packing Co.
Port Norris, NJ

Midshipman
Annapolis Canning Co.
Annapolis, MD

Miles
Miles, J.H. & Co.
Norfolk, VA

Miles River
Harrison & Jarboe Seafood Co.
St. Michaels, MD

MOCO
Milbourne Oyster Co.
Crisfield, MD

Monitor
Caldwell Bros. & Co.
Baltimore, MD

Monkey
Faet & Winebrenner
Baltimore, MD

Montauk
Montauk Seafood Co.
New York, NY

Moonlight Bay
Webb, E.I. & Co.
Weems, VA

National
National Fish & Oyster Co.
Baltimore, MD

Naumann
Naumann, R.E. Sea Food
Senora, VA

Navy
McCready, J.L. & Co.
Baltimore, MD

Negro Head
Auginbaugh Canning Co.
Biloxi, MS

Nelson's
Bon Secur Fisheries
Bon Secur, AL

New England
Warren Oyster Co.
Warren, RI

New Jersey
McConnell, Geo. A.
Port Norris, NJ

Nigger Head
Auginbaugh Canning Co.
Biloxi, MS

North Crystal
Brown, R.F. Sea Food Co.
Lansing, MI

North Crystal
Tignor Oyster Co.
Bowers, DE

NorVa
Unknown

Oasis
Battery Park Fish & Oyster Co.
Battery Park, VA

Ocean
Ocean City Oyster Co.
Ocean City, MD

Ocean Breeze
Haywood Oyster Co.
Perrin, VA

Ocean Cove
Bull, L.D.
Townsend, VA

Ocean View
Miles, J.H. & Company
Norfolk, VA

Oceanspra
Battery Park Fish & Oyster Co.
Battery Park, VA

Old Dominion
Irvington Packing Co.
Whitestone, VA

Old Dominion
Virginia Packing Co.
Baltimore, MD

Old Reliable
Lord-Mott Co.
Baltimore, MD

Old Salt
Tilghman Packing Co.
Tilghman, MD

Old Virginia
Cuthbert & Hughes, Inc.
Bohannon, VA

Olympia
Leonard, I.L. & Co.
Cambridge, MD

Originalpac
Loockerman, C.A.
Crisfield, MD

Quality
Newcomb & Hollinger Oyster Co.
Port Norris, NJ

Oval
Booth, A.
Baltimore, MD

Owls Head
Moore & Brady
Baltimore, MD

Oxford
Jones, Thos. E. Co.
Oxford, MD

Oyster Boat
Delaware Seafood Co.
Philadelphia, PA

Oyster World
Oyster World Inc.
Weems, VA

Pacific Oysters
United Oyster
Seattle, WA

Palm
Southern Shellfish Co.
Harvey, LA

Park
Wellman-Peck & Co.
San Francisco, CA

Patuxent
Denton, Warren & Co.
Broomes Island, MD

Patuxent River
Copsey, Leonard
Oraville, MD

Pearl
McNasby Oyster Co.
Annapolis, MD

Pedigree
De Jean Packing Co.
Biloxi, MS

Peerless
Pearson, C.H. & Co.
Baltimore, MD

Pelican
Dunbar-Dukate Co.
New Orleans, LA

Pelican
Pelican Oyster & Fish Co.
New Orleans, La

Perch Creek
Smith, B.G. & Son
Sharps, VA

Philson
Secretary Oyster Co.
Secretary, MD

Pimlico
Keagle, Wm. A. & Son
Baltimore, MD

Pin Point
Varn, A.S. & Son
Baltimore, MD

Pioneer
Stansbury, James E.
Baltimore, MD

Pittman Bewdley
Pittman, W.R. & Sons
Somers, VA

Pittman-Bewdley
Pittman, W.R. & Sons
Lancaster, VA

Plum Tree Island
Quinn, M.F.
Hampton, VA

Plymouth Rock
Hubbard, Jas & Son
Baltimore, MD

Plymouth Rock
Standard Fish & Oyster Co.
Baltimore, MD

Pocahontas
Chase, W.L. & Co.
Norfolk, VA

Pocomoke Sound
Pocomoke Sound Oyster Co.
Onancock, VA

Potomac View
Davis, Charles E.
Wynne, MD

Premium
Gude, Wm. D. & Co.
Baltimore, MD

Presto
Franks Seafood
Crisfield, MD

Pride
McClain, Wm. M.
Philadelphia, PA

Pride of Chesapeake
Dryden, Carol & Co.
Crisfield, MD

Pride of Chesapeake Bay
Leonard,John H.
Baltimore, MD

Pride of Gulf
Caernarvon Canning Co.
Caernarvon, LA

Pride of Maryland
Reddish, Charles W.
Baltimore, MD

Pride of the Chesapeake
Groves, J.D. & Co.
Baltimore, MD

Purebay
Rooks, B.J. & Son
Warren, RI

Purity
Ocean Seafood Co.
Baltimore, MD

Quality
Carson, L.R.
Crisfield, MD

Quality
Cedar-Rapids Commission Co.
Cedar Rapids, IA

Quality First
Elsworth, J & J Co.
Greenport, NY

Queen's Choice
B & S Fisheries Inc.
Fall River, MA

Queen's Choice
B & S Fisheries Inc.
Grasonville, MD

Randolph's
Randolph's Seafood
Coden, AL

Rappahannock River
Ferguson, J.W. Seafood Co.
Remlick, VA

Ray's
Ray's Seafood
Crisfield, MD

RCV
RCV Seafood Corp.
Morattico, VA

Red Cross
Elsworth, J & J Co.
Greenport, NY

Regina
Bower, F.C. & Co.
Baltimore, MD

Reliable
Reliable Seafood Co.
Crisfield, MD

Rig Co.
Riggin, W.E. & Co.
Crisfield, MD

River's Delight
Hogge, J.W. Seafood Co.
Hayes, VA

Riverside
Unknown

Rockaway
Hall Luhrs Co.
Sacramento, CA

Roland White's
White, Roland Oyster Co.
St. Bernard, LA

Rose Bay
Jennette, C.B.
Swan Quarter, NC

Rowe's
Rowe, H.C. & Co.
New Haven, CT

Rowley
Rowley Packing Co.
Chincoteague, VA

Royal
Peterson, G.F. & Co.
Baltimore, MD

S & M
Savage & Mears, Inc.
Chincoteague, VA

S & S
Sterling & Somers
Crisfield, MD

Saddle Rock
Store & Bunnel
Baltimore, MD

Saddle Rock
Tillman & Bendel, Inc.
San Francisco, CA

Sailor Boy
Brown, R.F. Sea Food Co.
Lansing, MI

Sailor Boy
East, F.F.

Maurice River, NJ

Sailor Girl
Plitt, Edwin M. & Son
Chicago, IL

Salt Air
Ranagan, Ralph H.
Boston, MA

Saltesea
American Oyster Co.
Providence, RI

Sandy Point
Davis, W.D. & Son
Wachapreague, VA

Satisfaction
Salisbury Packing Co., The
Salisbury, MD

Schwahn's
Schwahn, A.F. & Sons
Eau Claire, Ws

Scots Point
Olympia Oyster Co.
Shelton, WA

Sea "C"
Coulbourn Bros. Co.
Baltimore, MD

Sea Acre
Narragannsett Bay Oyster Co.
Providence, RI

Sea Acre
Narragannsett Bay Oyster Co.
Warren, RI

Sea Breeze
Texas Shrimp & Oyster Co.
Palacios, TX

Sea Crest
Roaring Point Oyster Co.
Nanticoke, MD

Sea Garden
Crosby, W.J., & Co.
Norfolk, VA

Sea Gull
Triggs, Chas. W. Co.

Sea Tang
Leonard, John H.
Baltimore, MD

Sea X Cross
Thompson, Geo & Son
Greenport, NY

Sea-Fresh
American Crabmeat Co.
Boston, MA

Sea-Kist
Miles, J.H. & Co.
Norfolk, VA

Sea-L-Tite
Killian, W.H. & Co.
Baltimore, MD

Sea-lect
Amory, G.W. Jr.
Hampton, VA

Seacrest
Kennerly, H.B. & Son
Nanticoke, MD

Seal
McGee, W.H. & Co.
Baltimore, MD

Sealshipt
Bluepoints Co., Inc.
West Sayville, NY

Seaman
Seaman Fish Co.
Bayou La Batre, LA

Seapure
Lester & Toner, Inc.
Greenport, NY

SEAsa-weHAK
Piney Island Seafood, Inc.
Moraticco, VA

Seawanhaka
Oyster Bay Oyster Co.
Oyster Bay, NY

Seminole
Unknown
Baltimore, MD

Sewansecott
Terry, H.M. Co.
Willis Warf, VA

Shamrock
Morrison & McCluan
Pittsburgh, PA

Shelter Island
Shelter Island Oyster Co.
Greenport, NY

Shemper's
Shemper Seafood Co.
Biloxi, MS

Shoal Creek
Winterbottom, W.G. & Co.
Cambridge, MD

Shorter's
Shorter's Place, Inc.
Benedict, MD

Sieward's
Sieward, F.H. & Co.
Baltimore, MD

Signal
Underhill, J.J. Co.
Baltimore, MD

Silver Sea
Jarrell & Rea
Baltimore, MD

Silver Sea
Silver Sea Oyster, Inc.
White Stone, VA

Smiling Oyster
Olympia Oyster Co.
Shelton, WA

Smith's
Smith, E.A. & Co.
Baltimore, MD

Some Oysters
Central Fruit Co.
Mansfield, OH

Somerset's
Somerset Seafood Co.
Deal Island, MD

Sparrer
Sparrer, W.H. & G.B.
Hampton, VA

Stag
Planters Packing Co.
Baltimore, MD

Stag
Planters Trading Co.
Baltimore, MD

Star
Ellis, Wm. L. & Co.
Baltimore, MD

Star
Handy, John T. Co.
Crisfield, MD

Star
Unknown, possibly Ward Oyster Co.
Crisfield, MD

Stork
Thomas & Thompson
Grasonville, MD

Stork
Thompson, H.S. & Co.
Grasonville, MD

Strickler's
Strickler, R.W.
York, PA

Summer Girl
Lee, H.D. Mercantile Co.
Waterbury, CT

Sun
Leib Packing Co., The
Baltimore, MD

Sun Set
Dorgan & McPhillips Pkg. Corp.
Mobile, AL

Sunbonnet
Indianapolis Fancy Grocery Co.
Indianapolis, IN

Superior
McNaney Oyster Co.
Baltimore, MD

Superior
Newcomb & Hand
Dover, DE

Superior
Newcomb & Hand
Rehobeth, DE

Superlative
Mears, A.H.G.
Wachapreague, VA

Tasty
Waterview Packing Co.
Waterview, VA

Tawes'
Tawes, J.C. & Son
Crisfield, MD

Temptation
McCoy, James Co.
Peoria, IL

Thomas
Thomas, Geo. H.
Cincinatti, OH

3-D's
Drewer, H.V. & Son
Saxis, VA

Tiger
Platt & Co.
Baltimore, MD

Tilghman
Tilghman Packing Co.
Tilghman, MD

Tom's Cove
Birch, William V.
Chincoteague, VA

Tom's Cove
Bunting, Wm. C. Oyster Co.
Chincoteague, VA

Tom's Cove
Tom's Cove Oyster Co.
Chincoteague, VA

Tom's Cove
Watson, Ralph E. Oyster Co.
Chincoteague, VA

Treat from the Deep
Bivalve Packing Co.
Bivalve, NJ

Tred Avon
Sinclair, E. & Co.
Oxford, MD

Tred Avon River
Harris, A.B. & Co.
Oxford, MD

Triangle
Wentworth, O.E. & Co.
Baltimore, MD

Triumph
Roberts, Geo. M. & Co.
Baltimore, MD

Tube Rose
Mebus & Drescher
Sacramento, CA

Tyler's
Tyler, Lawrence & Co.
Crisfield, MD

Valliant's Delight
Valliant, Wm. H. & Bro.
Bellevue, MD

Veach
Clayton, J.M. Co.
Cambridge, MD

Vernon's
Haywood, V.A. Seafoods
Perrin, VA

Very Best
New Fisheries Co.
Cincinnati, OH

Victory
Boyle, John H. Co.
Baltimore, MD

Viking
Schacht Seafoods, Inc.
New York, NY

Virginia
Crosby, W.J. & Co.
Norfolk, VA

Virginia Capes
Fast Bros.
Hampton, VA

Virginia Sea
Virginia Seafood Exchange
Newport News, VA

W & A
Warwick & Ashburn
Weems, VA

Wachapreague
Mears & Powell
Wachapreague, VA

Wagner's
Wagner, Martin Co.
Baltimore, MD

Wahkonsa
Wahkonsa Packing Co.
Fort Dodge, IA

Walker's Seaside
Walker, J.C. Brothers, Inc.
Exmore, VA

Ward
Ward, W.E., Oyster Co.
Crisfield, MD

Ware River
Ware River Seafood Co.
Schley, VA

Webster's Best
Webster-Butterfield Co.
Baltimore, MD

West's
West Oyster Co.
Oyster, VA

Westbrook
Morgan, W.F. & Sons
Weems, VA

Westbrook
Tracy & Avery Co.
Mansfield, OH

White City
Kunin, Samuel & Sons
Chicago, IL

White Rock
Lovejoy, Frederick F.
East Norwalk, CT

Wild Duck
Roberts, R.E. Co.
Baltimore, MD

Wilkerson's
Wilkerson, Herbert & Son
Colonial Beach, VA

Wilkerson's
Wilkerson, Herbert & Son
Potomac Beach, VA

Willapoint
United Oyster Producer's Assoc.
Seattle, WA

Williamsburg
York River Oyster Corp.
Glouster, VA

Willis
Willis Bros., Inc.
Williston, NC

Windmill
Paxton & Gallagher Co.
Omaha, NE

Winstead
Weems Seafood Co.
Weems, VA

Wishard's Oysters
Wisherd, D.N. & Sons
Oyster, VA

Woodfield's
Woodfield Fish & Oyster Co.
Galesville, MD

'XC'LENT
Collison, J.H., Co.
Baltimore, MD

York
Cook's Oyster Co.
York, VA

Index

including oyster packing companies and general subjects.

Specific brands of oysters are found alphabetically in the Oyster Brands list on pages 161-164. There, the associated packing company is given, which then can be located through this index.

a

b

c

S

t

u

V

W

x, y & z